This book is to be returned on or before
the last date stamped below.

A

- 7 APR 1976

12. DEC 84.

- 2 NOV 2010

A/612.0144

Donagh O'Malley,
R. T. C.
Letterkenny

# Physiology, Environment, and Man

# ENVIRONMENTAL SCIENCES

An Interdisciplinary Monograph Series

EDITORS

**DOUGLAS H. K. LEE**
National Institute of
Environmental Health Sciences
Research Triangle Park
North Carolina

**E. WENDELL HEWSON**
Department of Meteorology
and Oceanography
The University of Michigan
Ann Arbor, Michigan

**C. FRED GURNHAM**
Department of Environmental
Engineering, Illinois Institute of
Technology, Chicago, Illinois

ARTHUR C. STERN, editor, AIR POLLUTION, Second Edition. In three volumes, 1968

**DOUGLAS H. K. LEE**
National Institute of
Environmental Health Sciences
Research Triangle Park
North Carolina

**E. WENDELL HEWSON**
Department of Meteorology
and Oceanography
The University of Michigan
Ann Arbor, Michigan

**DANIEL OKUN**
University of North Carolina
Department of Environmental
Sciences and Engineering
Chapel Hill, North Carolina

L. FISHBEIN, W. G. FLAMM, and H. L. FALK, CHEMICAL MUTAGENS: Environmental Effects on Biological Systems, 1970

DOUGLAS H. K. LEE and DAVID MINARD, editors, PHYSIOLOGY, ENVIRONMENT, and MAN, 1970

*In preparation*

R. E. MUNN, BIOMETEOROLOGICAL METHODS

KARL D. KRYTER, THE EFFECTS AND MEASUREMENT OF NOISE

# PHYSIOLOGY, ENVIRONMENT, AND MAN

*Based on a Symposium Conducted by*
*The National Academy of Sciences–National Research Council*
*August, 1966*

EDITED BY

## DOUGLAS H. K. LEE and DAVID MINARD

NATIONAL INSTITUTE OF ENVIRONMENTAL
HEALTH SCIENCES
RESEARCH TRIANGLE PARK, NORTH CAROLINA

GRADUATE SCHOOL OF PUBLIC HEALTH
UNIVERSITY OF PITTSBURGH
PITTSBURGH, PENNSYLVANIA

### NAS–NRC COMMITTEE ON ENVIRONMENTAL PHYSIOLOGY

ANNA M. BAETJER
HARWOOD S. BELDING
BERNARD B. BRODIE
LOUIS S. GOODMAN

JAMES D. HARDY
STEVEN M. HORVATH
DAVID MINARD (CHAIRMAN)
NORTON NELSON

VAN R. POTTER
FREDERICK SARGENT
JAMES H. STERNER
JAMES L. WHITTENBERGER

ACADEMIC PRESS   New York and London      1970

COPYRIGHT © 1970, BY ACADEMIC PRESS, INC.
ALL RIGHTS RESERVED
NO PART OF THIS BOOK MAY BE REPRODUCED IN ANY FORM,
BY PHOTOSTAT, MICROFILM, RETRIEVAL SYSTEM, OR ANY
OTHER MEANS, WITHOUT WRITTEN PERMISSION FROM
THE PUBLISHERS.

ACADEMIC PRESS, INC.
111 Fifth Avenue, New York, New York 10003

*United Kingdom Edition published by*
ACADEMIC PRESS, INC. (LONDON) LTD.
Berkeley Square House, London W1X 6BA

LIBRARY OF CONGRESS CATALOG CARD NUMBER: 78-127690

PRINTED IN THE UNITED STATES OF AMERICA

*As they go to press, the Editors
learn with great regret of the
death of Colonel Adam Rapalski,
Division of Medical Sciences,
NAS–NRC, who labored long and hard
to bring together the presentations
to the Symposium into a coherent
and instructive document.
This volume is offered as a memorial
to his interest and skill.*

# Contents

*Foreword* . . . . . . . . . . . . . . . . xiii

**The Bretton Woods Symposium: Physiological Characterization of Health Hazards in Man's Environment** . . . . . . . . . 1
    DAVID MINARD

**Flow of Environmental Agents in Reaching Their Site of Action**
    LEWIS S. SCHANKER

| | |
|---|---:|
| General Principles of Membrane Penetration | 6 |
| Penetration of Membranes by Weak Electrolytes | 8 |
| Absorption from the Gastrointestinal Tract | 8 |
| Absorption from the Skin | 9 |
| Absorption from the Respiratory Tract | 9 |
| Penetration into Red Cells and Platelets | 9 |
| Penetration into the Central Nervous System | 10 |
| Liver | 12 |
| Kidney | 12 |
| Discussion and Conclusions | 12 |
| References | 14 |

**The Metabolic Fate of Common Environmental Agents** . . . . 15
    R. T. WILLIAMS

**Accumulation of Environmental Agents or Their Effects in the Body** . . 16
    ERIK WESTERMANN

References . . . . . . . . . . . . . . . . 26

**Interaction of Environmental Agents and Drugs** . . . . . . . 28
   JOHN J. BURNS

References . . . . . . . . . . . . . . . . 35

**Difficulties in Extrapolating the Results of Toxicity Studies in Laboratory Animals to Man** . . . . . . . . . . . 36
   DAVID P. RALL

References . . . . . . . . . . . . . . . . 42

**Some Prospects in Toxicology** . . . . . . . . . . 44
   BERNARD B. BRODIE

**Effects of Environmental Agents at the Genome Level** . . . . 49
   R. K. BOUTWELL

*Commentary by P. N. Magee* . . . . . . . . . . . 50

**Effects of Environmental Agents at the Level of Enzyme-Forming Systems**
   EMMANUEL FARBER

Aflatoxins . . . . . . . . . . . . . . . . 52
Carbon Tetrachloride . . . . . . . . . . . . . 54
Ethionine . . . . . . . . . . . . . . . . 55
Conclusion . . . . . . . . . . . . . . . . 56
References . . . . . . . . . . . . . . . . 56
*Commentary by A. H. Conney* . . . . . . . . . . . 57

**Effects of Environmental Agents at the Enzyme Levels— Air Pollutants** . . . . . . . . . . . . . . 59
   J. BRIAN MUDD

*Commentary by James R. Gillette* . . . . . . . . . . 60

## Growth and Trophic Factors in Carcinogenesis
### P. N. Magee

| | |
|---|---|
| Spontaneous Regression of Human Cancer | 61 |
| Dose-Response Relationships in Carcinogenesis | 62 |
| Two-Stage Theory of Carcinogenesis; Tumor Progression; Dependent and Autonomous Tumors | 62 |
| Hormonal Factors in Carcinogenesis | 66 |
| Immunologic Factors in Carcinogenesis | 72 |
| Nutritional Factors in Carcinogenesis | 73 |
| General Conclusions | 74 |
| References | 75 |
| *Commentary by Emmanuel Farber* | 77 |

## The Mechanism of Some Structural Alterations of the Lung Caused by Environmental Stresses . . . . . . . . . . . . 78
### Paul Gross

*Commentary by Hollis G. Boren* . . . . . . . . . . . . 79

## Mechanism of Bronchial Response to Inhalants
### Arthur B. DuBois

| | |
|---|---|
| Current Knowledge and Review of Some of the Literature | 80 |
| State of the Art | 82 |
| Speculations concerning Implications | 83 |
| References | 84 |
| *Commentary by Hollis G. Boren* | 87 |

## Principles and General Concepts of Adaptation
### C. Ladd Prosser

| | |
|---|---|
| Goals of Environmental Physiology | 89 |
| Criteria of Physiologic Variation | 90 |
| Molecular Mechanisms of Adaptation | 93 |
| Application of the Principles to Man | 95 |
| References | 99 |

## Human Genetic Adaptation . . . . . . . . . . . . 101
### C. C. Li

*Commentary by Bernard L. Strehler* . . . . . . . . . . . . 102

## Adaptive Cycles . . . . . . . . . . . . . . 103
Jürgen Aschoff

## Environmental Factors in Aging and Mortality
Bernard L. Strehler

I. Background and Definitions . . . . . . . . . . . 105
II. Categorical Analysis of Extrinsic Factors Affecting the Rate of Aging . . . 111
III. On the Possible Modification of the Rate of Aging through Manipulation
of the Internal Environment . . . . . . . . . . . 128
References . . . . . . . . . . . . . . . . 142
Commentary by Marott F. Sinex . . . . . . . . . . . 147

## Ecologic and Ethnic Adaptations
J. A. Hildes

Adaptation to the Physical Environment . . . . . . . . . 150
Biologic Factors in the Environment . . . . . . . . . . 152
Cultural Aspects of the Environment . . . . . . . . . . 152
Final Remarks . . . . . . . . . . . . . . . 153
References . . . . . . . . . . . . . . . . 153
Commentary by Cyril H. Wyndham . . . . . . . . . . 154

## Ecological Implications of Individuality in the Context of the Concept of Adaptive Strategy . . . . . . . . . . . . 155
Frederick Sargent

Commentary by A. Pharo Gagge and George Z. Williams . . . . . . 157

## Cross-Adaptation . . . . . . . . . . . . . 158
Henry B. Hale

References . . . . . . . . . . . . . . . . 165

## Comments on Cross-Adaptation
Melvin J. Fregly

Classification of Adaptations . . . . . . . . . . . 170
Cross-Adaptation . . . . . . . . . . . . . . 171
Methods for Study of Cross-Adaptation or Cross-Acclimation . . . . . 172
References . . . . . . . . . . . . . . . . 175

## Adaptation to Heat and Cold
CYRIL H. WYNDHAM

Adaptation to Heat . . . . . . . . . . . . . . . 177
Adaptation to Cold . . . . . . . . . . . . . . . 202
References . . . . . . . . . . . . . . . . . 203
Commentary by Harwood S. Belding . . . . . . . . . . 205

## Cardiac Disease in the Context of the Future Environment . . . . 206
STEVEN M. HORVATH

References . . . . . . . . . . . . . . . . . 210

## Adaptation and Environmental Control . . . . . . . . . 212
JAMES D. HARDY AND J. A. J. STOLWIJK

## Review and Comment on "Waste Management and Control"— A Report to the Federal Council for Science and Technology . . . 214
DONALD H. PACK

## How Is an Optimum Environment Defined?
VAN R. POTTER

Culture as Environment . . . . . . . . . . . . . 216
Physiologic Adaptation . . . . . . . . . . . . . 218
Enzyme Adaptation to Toxic Hazards . . . . . . . . . 219
Enzyme Adaptation to Daily Regimens . . . . . . . . . 220
Enzyme Adaptation to Fasting . . . . . . . . . . . 221
Definition of Optimum Environment . . . . . . . . . 224
References . . . . . . . . . . . . . . . . . 225
Additional References . . . . . . . . . . . . . 227

Index . . . . . . . . . . . . . . . . . . 229

# Foreword

*"Man does not limit himself to seeing; he thinks and insists on learning the meaning of the phenomena whose existence has been revealed to him by observation. So he reasons, compares facts, puts questions to them, and by the answers which he extracts, tests one by another."*

CLAUDE BERNARD, 1865\*

A century later, man's observations tell him with increasing insistence that all is not well with his environment; and for these disturbing phenomena he demands explanation—explanation of how they have come about, understanding of their effect on living matter, and assessment of what they ultimately mean for his cherished way of life.

The burgeoning resources of contemporary science have responded to this demand, not so much by the creation of new disciplines, as by the orientation of existing disciplines to questions of environmental moment. New names have been coined, old names compounded, and boundaries modified, but these are no more than superficial indicators of deeper changes. Academic science, devoted to the establishment of basic truths free from the pressures of "practical" problems, must and will continue to flourish in quiet corners; but the deployment of knowledge and the establishment of techniques for its deployment in the service of environmental management have been clearly recognized as additional and urgent needs. Fortunately, the exponential growth of science and technology, which has contributed—it must be admitted—to many of our environmental woes, has also developed the knowledge, techniques, and the individuals which can be used in their amelioration.

The term "environmental sciences" has been given to the body of knowledge that is relevant to problems of environmental effects and their management. Its boundaries are purely pragmatic. Any activity which subscribes to the tenets of scientific procedure and can contribute to the understanding of environmental problems qualifies, *ipso facto*, for inclusion.

It was with these thoughts that the Environmental Sciences Series of interdisciplinary monographs was established:

> The realignment of the biological, physical, and engineering sciences into a new field of Environmental Sciences had resulted from the rapid technological developments that have occurred in response to the urgent needs

---

\* "An Introduction to the Study of Experimental Medicine" (H. C. Greene, transl.), p. 5. Dover, New York, 1957. Reprinted through permission of the publisher.

of our expanding urban and industrial society. The Editors hope that this series will substantially influence and give direction to as yet unexplored areas of study as well as serve as a medium for the results of current research. (Editorial Policy)

Where effects on man are of primary concern, the term "environmental health sciences" is often used; but the fact that man is but part of an ecosystem—albeit a dominant and disturbing part—cannot be forgotten.

The central theme of the series is man in relation to his physical environment. However, because there is an obvious mutual interdependence between man and the plant and animal life around him, the interaction of environmental factors with other forms of life will be an integral part of this publication program. (Editorial Policy)

The Environmental Sciences Series was instituted to meet a need. Worthy material was already in preparation. The Second Edition of "Air Pollution," edited by Arthur C. Stern, in three volumes, was chosen as the inaugural item. The proceedings of the Bretton Woods Symposium on "Physiological Characterization of Health Hazards in Man's Environment," held in the fall of 1966, was chosen as the second entry in the series. These proceedings have appeared, in part, in the October number (Nos. 5–6, Vol. 2, 1969) of the periodical *Environmental Research*. They are reproduced here, together with synopses of other papers that were presented but not included in the journal, and summaries of formal discussion. Specially prepared monographs such as "Chemical Mutagens: Environmental Effects on Biological Systems" by L. Fishbein, H. Falk, and G. Flamm, now ready for publication, will continue the Series.

As Minard remarks in the opening chapter of the current volume, the National Academy of Sciences–National Research Council was asked to "undertake a broad-based critical study of the physiological underpinning of current concepts of biological responses to toxic chemicals and physical stresses." The nature of the participation is indicative of the extent to which even recent attitudes have changed in the biological sciences. The demands of World War II left environmental physiology with the definite image of being concerned primarily with the response of the total organism (man) to specific stresses which were predominantly physical. Appeal was made to organ systems for explanation of the responses observed, but seldom was cellular, let alone subcellular function considered. Less than one-third of the material presented at the Symposium, by even the most generous estimate, was of this character. The majority of the presentations, seen apart from the meeting's title, would most likely be labeled "molecular biology," or at least applications of that currently popular field. To a certain extent the emphasis on intracellular phenomena might be attributed to the fact that toxic chemicals were specifically cited in the assignment, but even with physical factors, such as heat, and certainly ionizing radiation, the ultimate response is now sought at the molecular, or at least at the organelle, level. New concepts, new techniques, and new types of investigator are involved, emphasizing, if any emphasis is needed, the futility of setting any boundaries on environmental sciences other than the pragmatic one of relevance.

While one might expect a textbook to present its field in organized and comprehensive fashion, a symposium necessarily follows more of an illustrative pattern, according to the personal interests or even idiosyncrasies of the participants. It is interesting to note that, in spite of these limitations, the presentations did in fact cover the range of physiological concerns with environmental effects, from the genetic to the temporal, and from the molecular to the holistic. This volume presents in microcosm the illustrative, or even opportunistic selectivity that the Series will revel in its macrocosm. In so doing, both this volume and the Series happily parallel a growing trend in higher education—that of developing individual capabilities around suitable topics, instead of trying to give every student a standardized, comprehensive and perchance indigestible concentrate of what authority thinks he ought to have. The success of the Series will, we hope, be judged by the success with which its readers apply the principles presented.

DOUGLAS H. K. LEE
E. WENDELL HEWSON
DANIEL A. OKUN
*Series Editors*

# The Bretton Woods Symposium: Physiological Characterization of Health Hazards in Man's Environment

DAVID MINARD

*Graduate School of Public Health, University of Pittsburgh, Pittsburgh, Pennsylvania*

Industrial growth and innovation based on scientific and engineering achievements are the means by which a technological society converts its resources into products and services which both enhance the material wealth of the nation and also provide sophisticated weaponry for its defense.

In the main, such growth raises living standards by increasing the quality and quantity of food, housing, transportation, and sanitation. By reducing toil, technology allows more leisure time for the citizens of a civilized society to enjoy educational and recreational pursuits.

On the other hand, technological progress coupled with population growth and urbanization, has introduced into the environment man-made hazards to health in the form of synthetic industrial and agricultural chemicals, toxic elements, industrial and community waste products, new sources and kinds of energy, as well as psychophysiological stresses such as crowding and noise. Detrimental effects on health of some of these stresses are immediately apparent, while other long-range effects are only dimly perceived, but considered by many as being of ominous dimensions.[1]

It is clear that our society must weigh carefully the benefits it enjoys from technological accomplishments and industrial growth against the costs it may have to pay in terms of present or future health impairment of its people stemming from man-made hazards and also from the loss of esthetic qualities of living resulting from deleterious changes in the human ecosystem.

The best available advice from authorities in various fields of science and engineering, commerce, the health professions, and the social sciences will be needed to identify, characterize, and anticipate public health hazards in the environment. Only from such informed sources can our society reach wise decisions in formulating operational plans for controlling further proliferation of environmental health hazards, and where feasible, restoring a more favorable environment.

The United States Public Health Service has long recognized its responsibility for health protection of the population not only against infectious diseases, but also against chemical and physical hazards in the community and work environments. To this end the USPHS had established action programs for the study and control of health hazards in various categories of the environment,

[1] Joint House-Senate colloquium to discuss a national policy for the environment. Hearing before the Committee on Interior and Insular Affairs, Ninetieth Congress, July 17, 1968, No. 8.

including water supply, milk and food sanitation, sanitary engineering, air pollution, radiological health, and occupational health.

In its report to the Surgeon General in 1962, the Committee on Environmental Health Problems (better known as the Gross Committee) outlined future national needs in environmental health and recommended substantial expansion in manpower and facilities for intramural research and training, in order to provide technical and advisory services to the states, and also to expand extramural programs of training and research in each of these categorical areas of environment health.

A major concern of the Gross Committee, however, was the need for a far more detailed understanding than now exists of the complex interactions of man as a biological system with the multiple physical, chemical, biological, and psychosocial environmental stresses which impinge upon him individually and in social groups. The Gross Committee urged the organization within the USPHS of an Office of Environmental Health Sciences which would include biological, physical, and social scientists as well as mathematicians whose mission would be to study and conduct research on basic problems in environmental health of common interest to the several categorical environmental health agencies within USPHS. The intent of the recommendation was that the research studies supported by this office would be more basic, longer term, and more interdisciplinary than was possible under then existing action programs.

In 1966, Congress authorized creation of an Environmental Health Center[2] which was subsequently established within the National Institutes of Health with a National Center of Environmental Health Sciences in the Research Triangle in North Carolina, to conduct the intramural research programs together with additional funds to support extramural programs of research and training through project and center grants to universities.

Anticipating the founding of DEHS (now NIEHS), scientists within USPHS undertook a re-evaluation of the role that physiology would play in investigations of the toxic effects of environmental agents on human health and the interplay of these agents with other environmental stresses. Their conclusion was that environmental physiology in the past had not sufficiently clarified the multiple chains of cause and effect mechanisms which exist between the exposure of man to environmental agents and the integrated response of his biological systems at the molecular, cellular, and higher levels of organization.[3]

Hence, in 1965 the USPHS proposed that the National Academy of Sciences-National Research Council undertake a broad-based critical study of the physiological underpinning of current concepts of biological responses to toxic chemicals and physical stresses.

NAS-NRC had previously taken the lead in studies in man's management of his environment and had issued two substantive reports: "The Management of National Resources," and more recently "Waste Management and Control" (the Spilhaus Report), which together with a third report emanating from the President's Advisory Committee, "Restoring the Quality of our Environment"

[2] Now the National Institute of Environmental Health Sciences.
[3] D. H. K. Lee (background paper for participants in the Bretton Woods Symposium).

(The Tukey Report), formed a thoughtful and comprehensive treatment of the effects of technological trends and innovations, as well as industrial expansion, on the natural environment and their resulting impact on human ecology. Specific recommendations dealt with means for identifying, predicting, and controlling deleterious changes in environment insofar as these reflect themselves in impairment of human health.

In response to the proposal of the USPHS, the Committee on Environmental Physiology (Appendix A) of the Academy Research Council Division of Medical Sciences—the latter then under the chairmanship of R. Keith Cannan—organized the Symposium on Physiological Characterization of Health Hazards in Man's Environment. With funding from the USPHS, this 2½ day meeting was held in August, 1966, at Bretton Woods, New Hampshire.

Among the participants were prominent authorities from biomedical disciplines ranging from molecular biology to human ecology, and including toxicology, biochemistry, oncology, pathology, and human genetics, as well as several subspecialties of physiology.

In the opening plenary session, the Co-Chairmen of the Symposium, Dr. Cannan and Dr. Minard, outlined the objectives of the meeting and proposed a plan of attack to be undertaken by the assembled scientists in fulfilling the aims of the Symposium.

In contrasting the relative simplicity of protecting human health against specific agents of infection and nutritional deficiencies, which had been largely controlled through application of principles of public health and environmental sanitation, the Conference Chairmen underscored the complexity of environmental health problems now facing the nation. The participants were asked to consider and discuss perplexing questions such as these: How does one assess the long-term effect of chronic exposure to low concentrations of toxic agents acting singly or in concert with other chemical or physical agents? How does one identify susceptible subgroups within the population who may be at special risk because of age, sex, genetic background, or pre-existing physical impairments? To what extent is the aging process itself accelerated by exposure to harmful agents acting at concentrations too low to elicit clearly more obvious signs of toxicity? To what extent are diseases of unknown etiology, such as chronic degenerative diseases affecting the cardiovascular and respiratory systems, aggravated by exposure to environmental agents?

Attention was called to growing evidence that a significant proportion of malignancies may be initiated, or promoted, by environmental factors. Unlike the great epidemic diseases of the past, present diseases associated with development and aging of the human organism are of complex etiology; multiple factors are involved and doubtless in many cases those inherent in the hosts are of primary importance. Nonetheless, a persevering search must go on to identify and control harmful factors stemming from the environment.

Further questions were those relating to man's capacity to adapt to a changing environment. Dire predictions had been heard about the possibility of reaching some point of no return in the buildup of toxic agents in our environment. Before entertaining such suggestions seriously, one must recognize that man, like other free-living organisms, possesses an adaptive plasticity to environmental change, such that prolonged exposure to stresses may lead to an increase in tolerance. This tolerance we designate as acclimatization in the case of physical agents, such as heat and altitude, immunity in case of exposure to infectious organisms, or chemical resistance in case of exposure to drugs and toxic chemicals. The whole learning process, indeed, is a behavioral adaptation mediated through the central nervous system, which ensures the integrity of the individual in the face of environmental stresses.

Adaptive responses, moreover, occur not only over brief periods but over the life span of the individual and indeed, by genetic selection, over a period of generations.

In the course of the Bretton Woods Conference, these and other questions were freely discussed and well-established concepts vigorously challenged. Rather than attempting to present the state of the art in his special field, each principal speaker sought instead, by use of examples as models, to identify gaps in our knowledge and to point to weaknesses in existing concepts. Thus, subsequent discussion was channeled in such a way as to bring fresh insight from various related disciplines.

There were five formal sessions of the Symposium (Appendix B) each being organized and moderated by one or two members of the Committee on Environmental Physiology. Each session concentrated on effects of environmental agents acting on mechanisms at one level of biological organization. These ranged in succession from the molecular and subcellular level, through that of cells and organ systems including discussions of absorption, transportation, metabolism, excretion and mechanism of action of environmental agents, next to the level of the intact human organism and its responses to these agents and finally to the level of populations in which discussants used epidemiologic data as indices of the range in susceptibility and adaptability in population subgroups.

By focusing in turn on each of these levels of biological organization, participants from disciplines other than those of the principal speakers in a given session were stimulated to suggest new ideas and to propose novel experimental approaches aimed at illuminating areas of ignorance. Eventually, one may hope that investigations undertaken by interdisciplinary teams of scientists, such as those represented at the Symposium, will establish a unifying framework to bridge conceptual gaps between the various biomedical disciplines upon which environmental health science must be based.

The ultimate objective of environmental physiology, to repeat in essence what was stated earlier, is to develop a basis for a systematic physiologic approach to an understanding of the complex interactions between man and the multiple physical and chemical factors in his environment, and thus to provide public health agencies with the knowledge they need not only to control harmful agents in the environment but to enhance man's ability to adapt effectively to technologic and social changes.

The flavor and substance of the Symposium on Physiological Characterization of Health Hazards in Man's Environment proved to be a stimulating experience for all participants. At least the Symposium demonstrated that fruitful dialogue and provocative ideas can be generated in an interdisciplinary forum by establishing channels of communication and cross fertilization between different fields of scientific endeavor. It may be hoped, at most, that the Symposium will be viewed in retrospect as the first tentative step in establishing a sound scientific and conceptual foundation for the emerging science of environmental health.

The largest measure of credit must be shared by the individual members of the Committee on Environmental Physiology who planned each session and proved to be such skillful moderators. The Committee Chairman expresses his gratitude to the USPHS which proposed such a gathering and which provided generously for the funding.

# Flow of Environmental Agents in Reaching Their Site of Action

LEWIS S. SCHANKER[1]

*Laboratory of Chemical Pharmacology,
National Institutes of Health, Bethesda, Maryland*

In considering the flow of environmental agents in reaching their site of action, we are concerned mainly with the passage of substances across the various body membranes. Man is separated from his external chemical environment by three major membranes: the skin, the epithelium of the alimentary canal, and the epithelium of the respiratory tract. Most substances can penetrate these boundaries, but the rates of penetration vary widely.

Once a substance has penetrated an external boundary, it enters the circulation and is carried throughout the body in one or more of several forms: as freely diffusible molecules dissolved in the plasma water; as molecules reversibly bound to proteins, chylomicrons, or other constituents of the serum; as free or bound molecules contained within erythrocytes and other formed elements; or finally as molecules bound to the surfaces of the formed elements.

As a substance is brought to the various body tissues, it encounters new barriers. First, there are the membranes of the individual tissue cells; then, within these cells, the membranes surrounding the nucleus, mitochondria, and so forth. Such a succession of membranes presents a formidable barrier to the penetration of many environmental agents. Moreover, even if the internal membranes serve only to slow somewhat the rate of penetration of a particular agent, they still afford a degree of protection for the organism in that they prolong the exposure of the agent to sites of metabolic destruction and excretion.

## GENERAL PRINCIPLES OF MEMBRANE PENETRATION

The various ways in which substances move across biological membranes can be grouped under two general headings: passive transfer processes and specialized transport processes. The term "passive transfer" implies that the membrane behaves as an inert, solvent-pore boundary and that solutes cross the boundary by diffusing through the solvent regions, by diffusing through the pores, or by flowing with water through the pores. The term "specialized transport" implies that the membrane displays an active character, transporting the solute in a manner that cannot be explained in terms of the structure or solvent properties of the membranes.

Many substances penetrate living membranes by simple diffusion; that is, their rate of transfer is directly proportional to their concentration gradient across the

---

[1] Present address: University of Missouri, Kansas City, 5100 Rockhill Road, Kansas City, Missouri 64110.

membrane. Some of the substances move across the membrane as though it were a layer of lipoid material, the speed of passage being determined by the lipid-to-water partition coefficient of the substances. In contrast, a number of lipid-insoluble substances of relatively small molecular size diffuse across the membrane as though it were interspersed with small, water-filled channels or pores, the smaller molecules crossing more rapidly than the larger ones.

Compounds can also move across membranes by a process of filtration or hydrodynamic flow. Thus, when a hydrostatic or osmotic pressure difference exists across a membrane, water flows, in bulk, through the membrane pores, dragging with it any solute molecules smaller than the pores. As an example, the water that filters across the renal glomerular membrane is accompanied by all the solutes of plasma except the protein molecules.

Although passive transfer across a lipoid-pore boundary adequately describes the penetration of membranes by many foreign organic compounds, it does not explain the rapid penetration by certain organic and inorganic ions, and some large, lipid-insoluble molecules, such as the monosaccharides. The concept of membrane "carriers" has arisen as a tentative explanation for the peculiar permeability of cell membranes to these substances. Carriers are pictured as membrane components capable of forming a complex with the solute at one surface of the membrane; the complex moves across the membrane, the solute is released, and the carrier then returns to the original surface.

There appear to be two major types of carrier transport, and, as work progresses, it will probably be found that there are many variations of these types. The type generally referred to as "active transport" has the following characteristics that distinguish it from a "diffusion process": (1) The solute moves across the membrane against a concentration gradient; that is, from the solution of lower concentration to the one of higher concentration; or, if the solute is an ion, it moves against an electrochemical potential gradient. (2) The transport mechanism becomes saturated when the concentration of solute is raised high enough. (3) The process shows specificity for a particular type of chemical structure. (4) If two substances are transported by the same mechanism, one will competitively inhibit the transport of the other. (5) The transport mechanism is inhibited by substances which interfere with cell metabolism.

Other types of carrier transport—for example, facilitated diffusion and exchange diffusion—have most of these characteristics, but they do not transport a substance against a concentration gradient.

A radically different type of specialized transport, in which cells engulf small droplets of the extracellular fluid, is the process known as "pinocytosis." It is seen in amebas, as well as in tissue cells growing in culture, and there is electron microscopic evidence that it occurs in mammalian cells within the animal. Very little is known about the physiologic significance of pinocytosis. Although the process appears to operate too slowly to account for the rapid cellular uptake of many substances, it could account for the uptake of small amounts of protein and other macromolecules. Another poorly understood form of specialized transport is the process of phagocytosis, in which some cells are able to engulf particulate or colloidal material.

## PENETRATION OF MEMBRANES BY WEAK ELECTROLYTES

Many drugs and other environmental agents are weak acids or bases and exist in solution as a mixture of the ionized and unionized forms. Since the unionized moiety is usually lipid soluble, and the ionized moiety lipid insoluble, only the unionized substance will diffuse rapidly across a lipoid membrane. The proportion of drug in the unionized form depends on the dissociation constant of the compound and on the pH of the solution in which it is dissolved. Consequently, in considering the passage of a weak electrolyte across a membrane, it is important to know the dissociation constant of the substance, as well as the lipid-to-water partition ratio of its unionized form.

## ABSORPTION FROM THE GASTROINTESTINAL TRACT

Studies of the absorption of many drugs and other foreign organic compounds from the stomach, small intestine, and colon have revealed that the gastrointestinal epithelium has the properties of a lipoid membrane. Most drugs are readily absorbed in their unionized form, and very slowly absorbed in their ionized form. Moreover, with compounds that exist mainly as unionized molecules, the relative rates of absorption are directly related to the lipid-to-water partition ratios of the molecules.

The stomach is a significant site of absorption for many acidic and neutral compounds. For example, salicylates and barbiturates, which exist as unionized molecules in the acid gastric contents, are rapidly absorbed. In contrast, such basic drugs as the plant alkaloids and many amines, which exist largely as ions, are very slowly absorbed.

In the small intestine and colon, whose contents have a pH of 6–8, most weak acids and bases are at least partially unionized and are absorbed at rates related to their lipid solubilities. The slowest rates of absorption are seen with completely ionized drugs, such as quaternary ammonium compounds and sulfonic acids, and with lipid-insoluble molecules, such as sulfaguanidine and mannitol.

In contrast with the compounds absorbed by diffusion are several classes of substances that cross the intestinal epithelium predominantly by specialized, active transport processes. These include lipid-insoluble substances required by the organism—for example, some sugars, amino acids, pyrimidines, and inorganic ions—and the fats, which exist in the intestine largely in the form of complex micelles. Foreign organic compounds can be absorbed by these active transport processes if their chemical structures are similar enough to that of the natural substrate. For example, the antitumor compounds 5-fluorouracil and 5-bromouracil are rapidly absorbed from the intestine by the process that transports natural pyrimidines, such as uracil and thymine. Moreover, foreign monosaccharides and amino acids have been shown to be actively absorbed from the small intestine.

Small amounts of macromolecules and particulate material are absorbed from the intestine by unknown mechanisms. For instance, resin particles as large as 5 $\mu$ in diameter are absorbed to a slight extent. Moreover, it is well known from the allergic response to some foods that small quantities of intact protein can cross the gastrointestinal epithelium. The toxicity produced on ingesting the exotoxin of *Clostridium botulinum* is another example of the absorption of trace amounts of a macromolecule. Perhaps these substances cross the intestinal boundary by pinocytosis or by leaking through tiny imperfections in the epithelium.

## ABSORPTION FROM THE SKIN

The results of numerous studies have indicated that organic compounds penetrate the skin predominantly by passing through a lipid-like barrier. This conclusion is based on many isolated observations that lipid-soluble molecules are absorbed much more readily than lipid-insoluble molecules and ions. Moreover, a study of the passage of nonelectrolytes across the excised rabbit skin has shown that various alcohols and urea derivatives diffuse across whole skin at rates roughly proportional to the ether-to-water partition coefficients of the compounds. The lipoid barrier of the skin appears to be within the epidermal layer, since the dermis is freely permeable to many solutes and displays the characteristics of a highly porous membrane.

For some time, there has been no general agreement among investigators as to the main pathway by which drugs traverse mammalian skin. Some authors have stressed the importance of the epidermal route, while others have contended that the appendageal route—through hair follicles, sweat glands, and sebaceous glands—is the predominant one. However, a recent study of the absorption of tri-n-butyl phosphate from the skin of living pigs has shown that the hair follicle is no more penetrable than an equivalent area of epidermis; in fact, regions of the skin devoid of hair follicles are penetrated slightly more rapidly than regions containing these structures.

## ABSORPTION FROM THE RESPIRATORY TRACT

Although it is well known that gases, as well as a variety of drugs administered as aerosols, readily penetrate the pulmonary membrane, almost nothing is known about the relative rates at which most substances are absorbed. It is possible that this boundary is more permeable than some of the other body membranes, since substances of low lipid solubility, such as some antibiotic compounds, appear to be absorbed fairly rapidly when administered as aerosols.

## PENETRATION INTO RED CELLS AND PLATELETS

After a substance has entered the bloodstream, it may then penetrate the red cells, platelets, and other formed elements.

A variety of basic organic compounds have been shown to enter the human red cell at rates related to their lipid-to-water partition coefficients at pH 7.4. For instance, compounds of high lipid solubility, such as aniline, procaine amide, and bufotenin, enter very rapidly; those of lower lipid solubility, like epinephrine, norepinephrine, and serotonin, enter considerably more slowly; and those of very low lipid solubility, such as tetramethylammonium and other quaternary amines, enter extremely slowly. Moreover, the cell membrane exhibits a perferential permeability to the unionized form of compounds; thus, when the pH of the extracellular fluid is varied, a basic compound like serotonin penetrates the cell at rates directly related to the proportion of compound present as the unionized moiety.

Organic acids have also been shown to penetrate the red cell at rates roughly parallel to their lipid solubilities; however, highly ionized acids of very low lipid solubility, such as phenol red, sulfanilic acid, and hippuric acid, diffuse into the cell much more rapidly than do highly ionized basic compounds of a similar low lipid solubility. Thus, the entry of organic anions resembles that of inorganic anions, which also penetrate the erythrocytes at rates greatly exceeding those of cations.

To explain the unusual permeability of the erythrocyte to anions, a number of workers have suggested that the cell is bounded by a lipid-like membrane which is perforated with positively charged aqueous channels of various diameters. According to this idea, small lipid-soluble molecules penetrate both by dissolving in the lipoid phase and by diffusing through the aqueous channels; larger lipid-soluble molecules enter mainly through the lipoid phase. Lipid-insoluble molecules, as well as anions, enter the cell if they are small enough to pass through the channels, whereas cations are largely excluded because of their lipid insolubility and their inability to pass through the positively charged channels. Perhaps large organic anions penetrate by way of the lipoid phase, with their charged portions protruding into the positively charged aqueous channels.

Although a number of organic substances attain higher concentrations within the red cell than in plasma, there is no evidence that this is a result of active transport into the cell. Rather, it has been demonstrated for several compounds that their localization within the cell results from binding to hemoglobin and perhaps other proteins.

There is evidence that the blood platelet is penetrated more readily by lipid-soluble substances than by lipid-insoluble substances. However, serotonin, a compound of low lipid solubility, has been shown to penetrate rapidly into the platelets of various species by an active transport process. Much more work is needed to characterize the permeability of this membrane-bounded structure.

## PENETRATION INTO THE CENTRAL NERVOUS SYSTEM

In passing from the plasma into the extracellular space of tissues, most solutes encounter little or no resistance at the capillary wall. Although the capillary endothelium does behave as a lipoid-pore boundary, the pores or intercellular spaces are so large that most substances penetrate readily.

An exception to this generalization is seen in the central nervous system, in which the capillary wall or the surrounding layer of glial cells forms a tight lipoid membrane that restricts the passage of many substances into the brain and cerebrospinal fluid (CSF).

As seen in Fig. 1, when various drugs and other organic compounds are administered intravenously to dogs, they diffuse from plasma into CSF at widely different rates. In general, compounds with high lipid-to-water partition ratios at pH 7.4, such as aniline, thiopental, and aminopyrine, enter very rapidly. But as the partition ratio falls off, the penetration rates decline markedly. A detailed comparison of the penetration rates with the lipid solubilities and degrees of ionization of the drugs has led to the following conclusions: (1) Lipid solubility is the rate-limiting factor with drugs that are mainly unionized in plasma; these

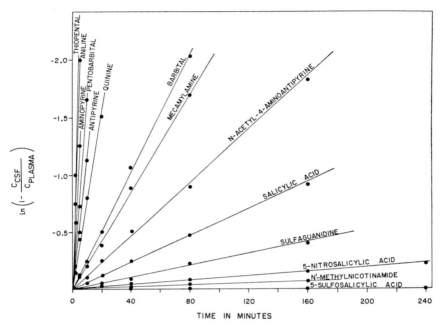

Fig. 1. Passage of drugs from the bloodstream into the cerebrospinal fluid (CSF) of dogs. $C_{CSF}$, concentration of drug in CSF; $C_{PLASMA}$, concentration of drug in plasma after correction for the degree of protein binding. The relative rates of penetration into CSF are given by the slopes of the lines.

compounds penetrate the blood-CSF boundary at rates related to the lipid-to-water partition coefficients of the unionized molecules. (2) The degree of ionization is the rate-limiting factor with compounds that are highly ionized in plasma; these drugs enter the CSF at rates roughly parallel to the proportion of drug unionized at pH 7.4. (3) Although both lipid solubility and the degree of ionization are important in governing the passage of drugs into CSF, lipid solubility is probably the dominant characteristic, since the relevance of the degree of ionization is probably a consequence of the poor lipid solubility of organic ions.

The passage of most organic compounds from plasma into brain tissue appears to be determined by these same principles. Some substances enter the brain and CSF by specialized carrier transport processes. Included in this group are glucose, some amino acids and inorganic ions, and probably a number of other natural substrates.

After a substance has entered the central nervous system, it returns to the bloodstream in two or three different ways: (1) by passive diffusion, (2) by filtration as the CSF drains into the blood through the large channels of the arachnoid villi, or (3) by active transport from CSF to blood across the epithelium of the choroid plexuses. The actively transported compounds include a number of highly ionized acids, such as hippurates, penicillin, and sulfonic acid dyes, and such ionized bases as hexamethonium, decamethonium, and $N^1$-methylnicotinamide. All the acids appear to be transported by one process and the bases by another process. Because of the active transport and filtration mechanisms that continually remove substances from the CSF, some compounds never attain a high concentration in the fluid when administered intravenously.

## LIVER

Although it has been shown that lipid-soluble compounds penetrate the liver more rapidly than do lipid-insoluble compounds, most substances enter the liver cell more readily than they do many other cells of the body, suggesting that liver cells have fairly large pores. In addition, the hepatic parenchymal cell has at least two active transport processes, one for a number of organic anions, and one for some quaternary ammonium cations. These polar substances are actively taken up by the cells and then secreted into the bile.

The anions secreted into bile include a wide variety of natural and foreign compounds, as well as glucuronides and other conjugates formed within the liver. When a lipid-soluble drug is conjugated with glucuronic acid, it becomes lipid-insoluble; the conjugate flows down the bile duct and enters the intestine. Within the intestine, the conjugate may be hydrolyzed to release the parent lipid-soluble drug, which is readily absorbed and returned to the liver for another cycle. Thus, some compounds tend to be retained in the body by the enterohepatic circulation.

## KIDNEY

In the kidney, solutes pass from plasma into the urine by filtration through the large pores of the glomerular membrane. As the urine moves along the renal tubule and becomes more concentrated, lipid-soluble substances are reabsorbed by diffusion across the tubular epithelium. In contrast, the lipid-insoluble substances are poorly reabsorbed and tend to be excreted.

In addition, the tubular cells have at least two active transport processes for the excretion of organic compounds: one for anions and one for cations. Many of these secreted substances are the same as those secreted by the liver and the choroid plexuses. The kidney also has active reabsorption mechanisms for sugars, amino acids, and perhaps other natural substrates.

## DISCUSSION AND CONCLUSIONS

It is important to recognize that, if a substance can diffuse rapidly across one of the body's external boundaries, it will most likely penetrate all subsequent boundaries with ease. This generalization follows from the fact that most biologic membranes have a number of properties in common—properties which determine their permeability characteristics. Thus, membranes, because of their lipoid nature, are highly permeable to lipid-soluble substances, provided that the substances are sufficiently water-soluble to approach the membrane in a molecularly dispersed form. Membranes are relatively impermeable to lipid-insoluble solutes, unless there are specialized carrier processes in the membrane for transporting the solutes. In addition, membranes are permeable to very small molecules, whether lipid-soluble or not, because the lipoid membrane appears to be interspersed with tiny water-filled channels or pores.

While the concept of the similarity of body membranes is important in understanding the absorption and physiologic distribution of many substances, it must be applied with caution because of two types of variation among membranes. One variation, about which little is known, concerns the size, shape, and electrical charge of membrane pores. The red cell, for instance, is highly permeable to anions and is accordingly thought to have positively charged pores; most other cells do not show this characteristic. The other variation concerns specialized transport processes, such as carrier-mediated transport, pinocytosis, and phagocytosis. Carrier processes, such as active transport and facilitated diffusion, generally bring about the rapid transfer of polar substances, which cannot readily diffuse across a lipoid membrane. Fortunately for the organism, the absorptive processes of this type are highly specific and are concerned mainly with bringing nutrients, such as sugars and amino acids, into the body and its cells and with regulating the intracellular concentration of inorganic ions. However, it is possible for a foreign substance to be taken into a cell by one of these specific processes if the chemical structure of the substance is similar enough to that of the natural substrate.

Carrier transport processes of rather low specificity have an important role in removing some toxic substances from the body. For example, many organic anions and cations—including drugs, their metabolites, and natural products of metabolism—are actively transported from CSF to blood, from blood into bile, and from blood into urine.

While body membranes are certainly of primary importance in regulating the flow of environmental agents through the organism, at least one other factor should be mentioned, and that is the binding of solutes to proteins and other macromolecules. The binding of most organic substances to biologic materials is reversible, involving the formation of relatively weak bonds, such as those of the van der Waals, hydrogen, and ionic types. However, some classes of substances—for example, the nitrogen mustards and the phosphorus-containing anticholinesterases—appear to form nonreversible, covalent bonds with cellular materials.

The reversible binding of a substance to plasma proteins helps to determine the concentration of freely diffusible substance in the various body tissues, since the tissue cells, being largely impermeable to protein, are in diffusion equilibrium with only the unbound fraction of the substance in plasma. If a second substance is introduced into the bloodstream, and if it displaces the first substance from its binding sites on the plasma proteins, the tissue levels of the first substance will rise in order to remain in equilibrium with the elevated concentration of unbound substance in plasma. With the increasing exposure of man to many drugs and other environmental agents, this type of competitive binding may be more common than presently realized. A normally safe dose of a drug may give dangerously high tissue levels if its usual degree of plasma binding is lowered by other agents.

From the point of view of environmental health, the most urgently needed information in the area of absorption, distribution, and excretion of substances is quantitative data on the absorption of environmental agents through the lungs and through the skin. There is a need for fundamental information on the precise ways in which substances cross these boundaries, and there is a need for information on the actual amounts of substances that pass from the environment into the body by these pathways.

Although we now have a fairly good understanding of how many substances are absorbed from the gastrointestinal tract, there are still wide gaps in our knowledge when it comes to some classes of environmental agents, including insecticides, organic ions, heavy metals, bacterial toxins, and other macromolecules. Once we have gained some understanding of the absorptive processes, there will remain numerous questions concerning the tissue distribution, protein binding, and excretion of environmental agents.

## REFERENCES

1. Brodie, B. B., and Hogben, C. A. M. (1957). Some physico-chemical factors in drug action. *J. Pharm. Pharmacol.* **9**, 345–380.
2. Dautrebande, L. (1962). "Microaerosols—Physiology, Pharmacology, Therapeutics. Academic Press, New York.
3. Davson, H. (1963). The cerebrospinal fluid. *Ergeb. Physiol. Biol. Chem. Exptl. Pharmakol.* **52**, 20–73.
4. Pappenheimer, J. R. (1953). Passage of molecules through capillary walls. *Physiol. Rev.* **33**, 387–423.
5. Peters, L. (1960). Renal tubular excretion of organic bases. *Pharmacol. Rev.* **12**, 1–35.
6. Schanker, L. S. (1962). Passage of drugs across body membranes. *Pharmacol. Rev.* **14**, 501–530.
7. Schanker, L. S. (1964). Physiological transport of drugs. *Advan. Drug Res.* **1**, 71–106.
8. Sperber, I. (1959). Secretion of organic anions in the formation of urine and bile. *Pharmacol. Rev.* **11**, 109–134.
9. Wilbrandt, W., and Rosenberg, T. (1961). The concept of carrier transport and its corollaries in pharmacology. *Pharmacol. Rev.* **13**, 109–183.
10. Wilson, K. (1961). New methods for the study of percutaneous absorption. *Drug Cosmetic Ind.* **88**, 444–529.

# The Metabolic Fate of Common Environmental Agents[1]

R. T. WILLIAMS

*Department of Biochemistry, St. Mary's Hospital Medical School,
London, England*

The gross pattern of the metabolic fate of most foreign chemical compounds can now be predicted fairly well, but the details are influenced by a variety of factors.

Two phases are usually involved. In Phase I, enzymes convert the compound by oxidation, reduction, or hydrolysis to compounds which can undergo Phase II. The Phase I reactions are very numerous and are frequently carried out by enzymes which occur in the liver microsomes. On occasion the products are more toxic than the parent compound (e.g., parathion). The Phase II reactions are few in number (seven of the nine most common reactions can occur in man) and are more widespread, occurring in liver mitochondria, intestine, and blood, as well as in liver microsomes. Only rarely do they fail to inactivate or detoxicate the compounds.

Excretory products may include: (1) the parent compound; (2) Phase I reaction products; and (3) Phase II reaction products. On the whole, the animal body has the biochemical equipment to deal with small quantities of a large number of foreign chemicals, but its limitations must be borne in mind when new chemicals are introduced into the environment for various technological reasons.

---

[1] Synopsis of paper sent to the Bretton Woods Symposium but not presented in person.

# Accumulation of Environmental Agents or Their Effects in the Body

ERIK WESTERMANN

*Department of Pharmacology, Medical School of Hannover, Hannover, Germany*

By definition, the environment is the aggregate of all external conditions and influences that affect life and development of an organism; therefore, environmental agents consist not only of toxic compounds emitted by heavy industry, but also of drugs used as pesticides or even in medical practice.

"Accumulation" merely implies that the rate of appearance is higher than the rate of disappearance. Therefore, environmental agents may accumulate either in the environment of man (air, water, food) or in the body of man. The question whether or not foreign compounds can accumulate in the body depends predominantly on their behavior in the organism. This is best illustrated in heavy smokers. Apart from various carcinogens, tobacco smoke contains nicotine and considerable amounts of carbon monoxide. Although both of these toxic agents are rapidly adsorbed by the respiratory tract, only carbon monoxide shows a distinct accumulative pattern; in other words, the concentration of carbon monoxide steadily increases in the body, whereas the concentration of nicotine does not. This difference in the behavior of the two compounds is due to their different rates of elimination. Nicotine is so rapidly metabolized in the liver and other organs that during the intervals of nonsmoking the concentration of this agent decreases to very low levels. Carbon monoxide, on the other hand, is so tightly bound to the hemoglobin of red blood cells that its elimination takes many hours. Therefore, the inhaled carbon monoxide is added to carbon monoxide not yet eliminated, thus further increasing its concentration in the blood. Carbon monoxide levels up to 10% have been measured in the blood of heavy smokers. An additional 5–8% from exposure to community air pollution may pose added risks to health.

Since the body of man is more than a simple solvent for foreign compounds, environmental agents may accumulate not only substantially, but also functionally in the body. In other words, there is sometimes a complete lack of correlation between drug concentration and drug effect. This different behavior can be illustrated with certain pesticides.

Chlorinated hydrocarbons, such as DDT, which are widely used as pesticides in agriculture, accumulate substantially in man. Because of their high fat solubility and low rate of elimination, these compounds are stored predominantly in the fat depots. Even the average American adult has accumulated more than 50 mg of DDT in his body. Fortunately, the toxicity of DDT is rather low in mammals. Toxic symptoms in man have been observed only at oral doses of 10–20 g of drug—enough to kill more than 1000 tons of flies.

Organophosphorus pesticides, such as parathion, on the other hand, are also very toxic in man. In contrast with the chlorinated hydrocarbons, these compounds do not accumulate substantially in the body, since they are rapidly metabolized. However, small daily doses show a distinct accumulative pattern, leading to clinical symptoms resembling strong cholinergic stimulation. This effect is due to a nonreversible blockade of cholinesterase, thus leading to an accumulation of endogenous acetylcholine at all cholinergic links in the organism. The accumulative pattern of parathion is best illustrated by experiments published in 1949 by Dubois and co-workers (Fig. 1).

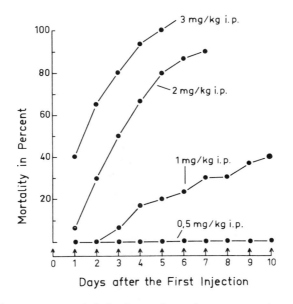

Fig. 1. Cumulative toxicity of daily doses of parathion in rats. (According to Du Bois et al. (2)).

Groups of rats received an intraperitoneal injection of parathion every day and the mortality was checked during a period of 10 days. At a daily dose of 0.5 mg/kg, none of the animals died during this period. If, however, the daily dose was increased to 1 mg/kg, mortality began at the third day; and after daily doses of 2 and 3 mg/kg, all the animals died within a few days.

In another experiment (Fig. 2) by Dubois and co-workers, cholinesterase was measured after daily administration of 1 mg/kg of parathion. It is interesting to note that the mortality in this group started when brain cholinesterase was blocked by more than 90% on the third day. Since the first dose of parathion already reduced cholinesterase activity by more than 50%, one would expect the second dose of the drug to block the enzyme completely. This is obviously not the case. As indicated by the broken lines, there was a partial recovery of enzyme activity during the 24-hour interval between the injections. Therefore, the inhibition which persisted more than 24 hours was added to the effect of the next

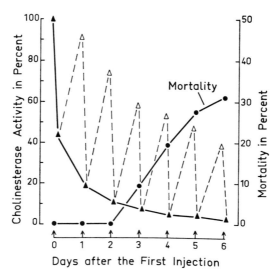

Fig. 2. Brain cholinesterase activity and mortality after daily injection of parathion (1 mg/kg ip) in rats. (According to Du Bois et al. (2)).

dose, thus slowly reducing the enzyme activity below the critical 10% level where mortality started.

When we tried to reproduce these experiments a few years ago, we obtained different results. In our hands, 1 mg/kg of parathion was not toxic at all: No mortality was observed even after daily intramuscular injection for 7 days. First, we thought that the negative result was due to strain differences, but later we found that the route of administration was the jumping point in the activity of parathion.

Also in mice, parathion (Table I) was much more toxic when injected intraperitoneally than when injected intramuscularly. This difference in activity was particularly striking at a dose of 5.6 mg/kg: After intraperitoneal injection, 86% of the animals died, while after intramuscular injection of the same dose, only 8% were killed. Cholinesterase in brain was also much more inhibited after intraperitoneal injection of parathion than after intramuscular injection.

TABLE I

Toxicity of Parathion after Intraperitoneal (ip) and Intramuscular (im) Injection in Mice

| Parathion (mg/kg) | Mortality in % after: | |
|---|---|---|
| | ip injection | im injection |
| 3.5 | 14 | 0 |
| 4.4 | 53 | 0 |
| 5.6 | 86 | 8 |
| 7.0 | 100 | 42 |
| 8.9 | 100 | 100 |

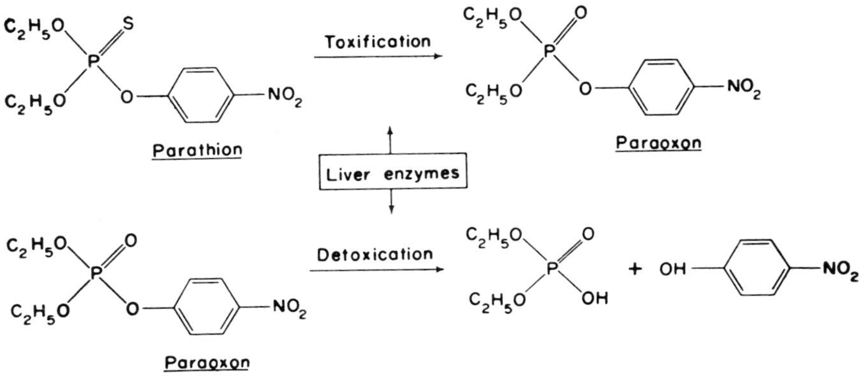

FIG. 3. Transformation of parathion.

It is well known that parathion itself is no inhibitor of cholinesterase, but is converted in the body to the active compound.

By replacement (Fig. 3) of the sulfur with an oxygen atom, parathion is converted to the highly active cholinesterase inhibitor paraoxon, a process which may be called "toxification," while the cleavage of the paraoxon molecule means "detoxication." The fact that parathion is more active when injected intraperitoneally means that its toxification by passing the liver—the conversion to paraoxon —occurs more rapidly than the detoxication of paraoxon formed.

What is the consequence of this long-lasting inhibition of cholinesterase? Since acetylcholine is the mediator of cholinergic neurons, one would expect at least a severe stimulation of the parasympathetic nervous system (for example, a fall in blood pressure and heart rate). However, the opposite is true!

Blockade of cholinesterase (Fig. 4) by paraoxon produced in rats a marked and long-lasting increase in blood pressure and heart rate. This effect is due to an excitation of cholinergic link in the brain, which in turn stimulates sympathetic centers. Paramacologic analysis has disclosed that the pressor effect of paraoxon is prevented not only by cholinergic blocking drugs, but also by central depressants, such as barbiturates (or mebutamate) and by peripheral blockade of the sympathetic nervous system. From these data, it can be concluded that the inhibition of brain cholinesterase stimulates vasomotor centers, thus leading to an

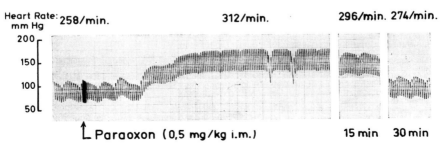

FIG. 4. Increase in heart rate and blood pressure following the injection of paraoxon (rat 185 g, urethane anesthesia).

TABLE II

Biochemical Changes Induced by Paraoxon (0.5 mg/kg im) in Rats

|  | Before paraoxon | After paraoxon | Change (%) |
|---|---|---|---|
| Blood glucose, mg/100 ml | 110 | 205 | +87 |
| Plasma free fatty acids, µeq/ml | 0.23 | 0.71 | +206 |
| Plasma corticosterone, µg/ml | 0.18 | 0.52 | +188 |
| Tryptophan-pyrrolase, µmoles/g/hr | 2.19 | 3.63 | +67 |
| Tyrosine-transaminase, µmoles/g/hr | 51.2 | 84.7 | +64 |

adrenergic pressor response. In addition, paraoxon produced a broad spectrum of biochemical changes (Table II).

As summarized in this table, the level of blood glucose was elevated by 100% and the level of plasma free fatty acids rose even by 200% within 2 hours after the injection of paraoxon. There is no doubt that hyperglycemia is the result of sympathetic stimulation induced by the cholinesterase inhibitor, and is mediated by a hypersecretion of catecholamines from the adrenal medulla. The increase of plasma free fatty acids, however, is probably not mediated by the sympathetic nervous system. We have been able to show that the lipolytic action of paraoxon was not prevented by adrenergic blockade, but was completely abolished by hypophysectomy. Obviously, inhibition of brain cholinesterase produces a discharge of lipolytically active pituitary hormones.

This view is further supported by the finding that paraoxon produced a rise in the level of plasma corticosterone and an increase in the activity of tryptophan-pyrrolase and tyrosine-transminase of the liver (Table II). The accelerated production of corticosterone is caused by a direct action of ACTH on the adrenal gland, which can be demonstrated also *in vitro* by incubating slices of adrenals with ACTH. The increased activity of the liver enzymes is a secondary response, mediated by the elevated level of plasma corticosterone. Since maximal activation of these liver enzymes is attained only after several hours of increased corticoid level, these indices are particularly useful in describing a prolonged pituitary-adrenal stimulation.

The mechanism of action of the pesticide paraoxon is a very specific one: It only produces a long-lasting blockade of cholinesterase, thus leading to an accumulation of endogenous acetylcholine. But, since acetylcholine is the mediator at all cholinergic links in the organism, it becomes clear why this drug interferes with so many biologic systems, followed by a broad spectrum of pharmacologic and biochemical effects.

Now I would like to discuss the mode of accumulation of some drugs which are frequently used in the treatment of psychiatric disorders. Both chlorpromazine and reserpine show a distinct accumulative pattern in animals and man when administered in daily small doses. However, in the case of chlorpromazine, there is a close relationship between the drug concentration in brain and its central depressant effect, while in the case of reserpine, there is a complete lack of correlation between drug level and drug effect. This different behavior is shown in Fig. 5.

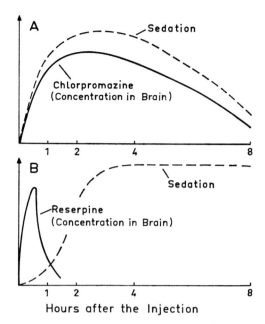

Fig. 5. Drug effect and drug concentration in brain. (According to Brodie and Costa (1)).

After injection of chlorpromazine (upper graph), the drug concentration in brain reaches a peak within 2 or 3 hours and then slowly declines. The central effect of chlorpromazine, sedation, also increases up to 3 hours after the injection and then slowly subsides. In other words, there is a positive correlation between drug effect and drug concentration in brain.

Reserpine (lower graph) also produces profound sedation maximally at about 2 or 3 hours after the injection. However, at this time nearly all of the drug has disappeared from the brain, while 15 minutes after the injection, when reserpine concentration in brain has reached maximal values, the drug has no sedative action at all. In this case, there is a complete lack of correlation between drug concentration and drug effect in brain. Nevertheless, reserpine also shows a distinct accumulative pattern when injected in daily small doses. Rats received 0.2 mg/kg of reserpine daily (Fig. 6). The behavior of the animals did not change until the third day. But shortly after the injection of the third dose of reserpine, profound sedation developed.

The work of Dr. Brodie (1) and his group has disclosed that the central effects of reserpine are associated with changes in the amine level of the brain. The first dose of reserpine lowered the brain amine level to about 75% of normal, but no sedation occurred. The second dose of reserpine reduced brain amine levels to about 55% of normal and still did not affect behavior. But the third dose, which lowered brain amines to about 30% of normal, elicited marked sedation. Although the drug does not accumulate substantially in the body, as does chlorpromazine, but is metabolized within a few hours after administration of each daily dose, the biochemical changes induced by reserpine are accumulative and closely related

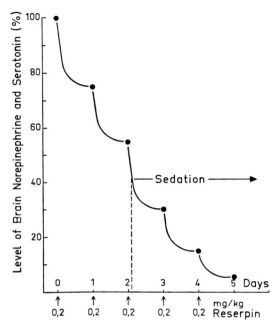

Fig. 6. Cumulative effects of daily small doses of reserpine in rats.

to its central effects. This conclusion is further supported by experiments with reserpine analogues.

Rats received a relatively high dose of various reserpine analogues (2.5 mg/kg), and brain amine level as well as behavior was checked 1 to 4 hours later (Table III). Only those compounds which depleted the brain of amines were found to produce sedation. Of particular significance is the finding that isoreserpine and isoraunescine, the stereoisomers of reserpine and raunescine, neither released brain amines nor produced sedation.

The mechanism underlying the effects of reserpine is considered to be a long-lasting impairment in the binding capacity for biogenic amines with subsequent depletion. Since all organs are depleted of amines—primarily norepinephrine—reserpine interferes with all funtcions in which adrenergic transmission is involved. Apart from its therapeutically desired hypotensive and sedative action, reserpine produces a broad spectrum of biochemical effects.

TABLE III
Responses to Various Reserpine Analogues (2.5 mg/kg iv)

| Compound | Brain amines | Behavior |
| --- | --- | --- |
| Reserpine | Depleted | Sedation |
| Isoreserpine | Normal | Normal |
| Raunescine | Depleted | Sedation |
| Isoraunescine | Normal | Normal |
| Rescinnamine | Depleted | Sedation |
| Serpentine | Normal | Normal |

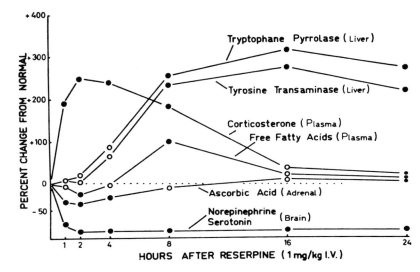

Fig. 7. Pituitary-adrenal responses of rats to reserpine.

Following a single dose of reserpine (Fig. 7), brain amine levels decreased and remained very low for more than 24 hours. In addition, the drug produced a sustained effect on various indices of pituitary-adrenocortical stimulation: a decrease in adrenal ascorbic acid, a rise in the level of plasma corticosterone, and an increase in the activity of tryptophan-pyrrolase and tyrosine-transaminase of the liver. These data clearly show that the reserpine induces a hypersecretion of ACTH which persists long after the drug has disappeared from the body and which is correlated with the long-lasting depletion of brain amines. However, while brain amine levels remained low for up to 48 hours, the plasma corticosterone level returned to normal values within 16 hours.

This dissociation finds its explanation in that the reserpine-induced hypersecretion of ACTH is followed by a decrease in the content of this pituitary hormone and a subsequent inhibition of pituitary-adrenal responses to stressful stimuli.

In normal rats, cold exposure produces a hypersecretion of ACTH as indicated (Fig. 8) by the elevation of plasma corticosterone up to 0.5 $\mu$g/ml. After pretreatment with reserpine (5 mg/kg, 24 hours previously), the animals were unable to respond to cold exposure with a hypersecretion of ACTH and the plasma corticosterone level remained practically unchanged. The failure of the reserpinized animals to respond to the stress with an elevation of plasma corticosterone was obviously due to a substantial depletion of ACTH in the pituitaries, which are already secreting the hormone as rapidly as possible to maintain a normal corticoid level, but which are unable to show an additional response to stressful stimuli.

Irrespective of how reserpine causes this excessive discharge of ACTH (by a depression of inhibitory hypothalamic pathways or by an activation of stimulatory fibers which liberate a corticotropine-releasing factor), an important clinical point

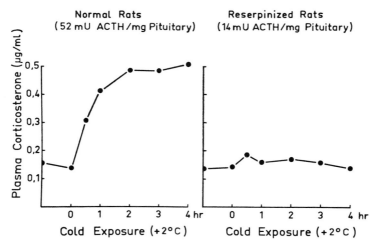

Fig. 8. Blockade of the pituitary-adrenal axis by reserpine (5 mg/kg iv).

should be raised. Since small doses of reserpine, given repeatedly, exert an accumulative effect also on brain-pituitary pathways, it is imperative to obtain more information on biochemical alterations induced by long-lasting treatment in man. Since the drug can impair the responsiveness of the pituitary-adrenal system, it is possible that reserpinized patients might require prophylactic supplementation with corticoids when they come under the stress of operation or injury.

Another example of so-called hit-and-run drugs are the monoamine oxidase (MAO) inhibitors, which are used in medical practice as psychic energizers and antidepressants. These agents are long-lasting because considerable time is needed for monoamine oxidase to recover or to be reformed. Given in small daily doses, these inhibitors inactivate a little of the enzyme each day. In this way, the inhibition of the enzyme gradually increases, even though the drug concentration in the tissues does not. Finally, when most of the enzyme is blocked, catecholamines and serotonin accumulate in the body and exert their pharmacologic actions. Blockade of MAO not only produces an accumulation of endogenous amines, but also destroys an important barrier which normally prevents an accumulation of exogenous amines.

In normal rats, the intramuscular injection of dopamine (10 mg/kg) produces a long-lasting increase in blood pressure, as shown in the upper curve in Fig. 9. If, however, the same dose of dopamine is injected intraperitoneally (middle curve), blood pressure remains unaffected. This difference in activity is due to the fact that, after intraperitoneal injection, a drug has to pass the portal system before it can enter the systemic circulation, while a drug injected intramuscularly enters the systemic circulation directly. Dopamine is an excellent substrate of MAO, and, during one single passage through the liver, the amine is metabolized to such an extent that only a small amount enters the systemic circulation—obviously too little to elicit pressor response. If, however, MAO is blocked (lower

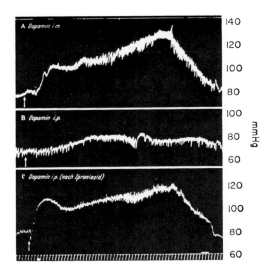

Fig. 9. Pressor effect of dopamine in rats after intramuscular (im) and intraperitoneal (ip) injection.

curve), the intraperitoneal injection of dopamine produces a sustained pressor effect.

Five years after we had published these results, it became apparent that this effect of MAO inhibitors is also of practical importance in that it increases the hazards associated with the consumption of normally innocuous food stuffs. In patients treated with MAO inhibitors, such as nialamide or pargyline, severe hypertensive reactions have been observed following ingestion of cheese or broad beans (*Vicia faber* L.). This effect also has been demonstrated in animal experiments.

In normal cats or rats, the feeding Emmenthal cheese or broad beans (Fig. 10) at a dose of 5 g/kg has no effect on the blood pressure. If, however, MAO is blocked by pretreatment of the animals with nialamide or pargyline, the administration of cheese or beans causes a marked increase in blood pressure and heart rate.

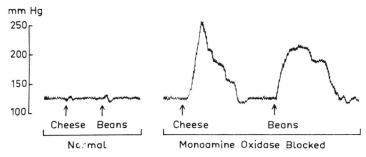

Fig. 10. Pressor response to intraduodenal administration of Emmenthal cheese and broad beans in the cat after blockade of monoamine oxidase. (According to Natoff (5) and Hodge et al. (3)).

It has been known for a long time that cheese contains relatively large amounts of tyramine. Actually, the name tyramine is a derivative of the Greek word "tyros," which means "cheese," and tyramine means "cheese amine." This amine is an excellent substrate of MAO, and, therefore, is rapidly metabolized in the gut and in the liver. If, however, the enzyme is blocked after treatment with nialamide, tyramine is able to pass the gut and the liver and to exert its pressor action.

Broad beans have a very similar mode of action. Two years ago, Hodge (3) and co-workers reported that the bean pods contain relatively large amounts of dopa—the precursor of dopamine and norepinephrine. Normally, the formed dopamine is rapidly metabolized by MAO, but, if the enzyme is blocked, the intact dopamine molecule can enter the systemic circulation and exert its pressor action. In this way harmless environmental agents can accumulate substantially in the body and produce severe toxic symptoms.

TABLE IV
Mode of Accumulation Various Substances

| Environmental agents | Mode of accumulation in the body | |
|---|---|---|
| | Substantial | Functional |
| Pesticides | DDT | Paraoxon |
| Central depressants | Chlorpromazine | Reserpine |
| Psychic energizers | Amphetamine | Nialamide |

Table IV summarizes the main points of my presentation. In the body of man, foreign compounds may accumulate substantially because of their low elimination rates. This behavior has been shown for the pesticide DDT, for the central depressant chlorpromazine, and for the psychic energizer amphetamine, an amine which is not metabolized by MAO.

However, there are compounds which are very rapidly eliminated from the body but nevertheless show a distinct accumulative pattern. This behavior, which might be called "functional accumulation," is brought about by a long-lasting blockade of important biologic mechanisms. Paraoxon, the active metabolite of the pesticide parathion, inactivates the enzyme cholinesterase; the central depressant reserpine interferes with the storage mechanism of biogenic amines; and nialamide inactivates the enzyme MAO.

Since these compounds interfere with the action of transmitter substances in the body, it becomes clear why a specific mechanism of a drug may impair the function of so many biologic systems and produce a broad spectrum of pharmacologic and biochemical effects.

REFERENCES

1. Brodie, B. B., and Costa, E. (1961). Some current views on brain monoamines. *Symp. Monoamines Systeme Nerveux Central, Bel Air, Geneva*, pp. 13–49. Georg. Geneva, Switzerland, 1962.
2. DuBois, K. P., Doull, J., Salerno, P. R., and Coon, J. M. (1949). Studies on the

toxicity and mechanism of action of *p*-nitrophenyl diethyl thionophosphate (Parathion). *J. Pharmacol. Exptl. Therap.* **93,** 79–91.
3. Hodge, J. V., Nye, E. R., and Emerson, G. W. (1964). Monoamine oxidase inhibitors, broad beans, and hypertension. *Lancet* **1,** 1108.
4. Holtz, P., and Westermann, E. (1959). Giftung und Entgiftung von Parathion und Paraoxon. *Arch. Exptl. Pathol. Pharmacol.* **237,** 211–221. Beeinflussung der Narkosedauer durch Hemmung der Cholinesterase des Gehirns. *Arch. Exptl. Pathol. Pharmacol.* **233,** 438–467 (1958). Psychic energizers and antidepressant drugs. *Physiol. Pharmacol.* **2,** 201–254 (1965). Academic Press, New York.
5. Natoff, I. L. (1964). Toxic manifestations of foodstuffs during drug therapy. *Proc. European Soc. Study Drug Toxicity* **4,** 158–166.
6. Stock, K., and Westermann, E. Untersuchungen über den Mechanismus der narkoseverkürzenden Wirkung von Monoaminooxydase-Hemmstoffen. *Arch. Exptl. Pathol. Pkarmacol.* **243,** 44–64 (1962). Über den Mechanismus der lipolytischen Wirkung des Physostigmins. *Arch. Exptl. Pathol. Pharmacol.* **252,** 334–354 (1966).
7. Westermann, E. Some metabolic factors controlling duration of drug action. *In* "Mode of Action of Drugs," Proc. Internat. Pharmacol. Meeting **6,** 205–211, Macmillan (Pergamon), New York (1962). Cumulative effects of reserpine on the pituitary-adrenocortical and sympathetic nervous system. *In* "Drugs and Enzymes," Proc. Internat. Pharmacol. Meeting **4,** 381–392, Macmillan (Pergamon), New York (1965).

# Interaction of Environmental Agents and Drugs[1]

JOHN J. BURNS

*Hoffmann-La Roche Inc., Nutley, New Jersey*

Exposure of animals to various environmental agents, such as drugs, pesticides, and carcinogenic polycyclic hydrocarbons (Table I), can stimulate the metabolism of drugs (5, 10, 11, 13, 15). The environmental chemicals exert this action by increasing the amount of drug-metabolizing enzymes in liver microsomes, which is referred to as enzyme induction. These enzymes metabolize many clinically useful drugs by various reactions, including N-dealkylation, deamination, hydroxylation, and glucuronide formation. Enzyme induction has considerable importance in pharmacologic and toxicologic studies in experimental animals, and recent work indicates that this may also hold true for drug therapy in man (1, 2).

The ability of drugs to increase the synthesis of drug-metabolizing enzymes in liver microsomes is a general phenomenon, since it occurs in the monkey, rat, mouse, guinea pig, dog, rabbit, and cat. Many different types of drugs have been shown to exert this action in animals—namely, barbiturates and other hypnotics, analgesics, tranquilizers, antihistaminics, oral antidiabetics, and uricosurics. Enzyme induction explains some unusual pharmacologic effects that occur when drugs are given in combination and accounts for tolerance observed on prolonged treatment with some drugs.

The liver microsomal enzyme that hydroxylates zoxazolamine (Fig. 1) is an example of a drug-metabolizing enzyme that is stimulated by drug administration (5). Zoxazolamine is hydroxylated in the 6-position by liver microsomes to give a pharmacologically inactive metabolite, so that any increase in the activity of this enzyme system would be expected to decrease the duration of action of zoxazolamine. The effect of phenobarbital and 3,4-benzpyrene on the zoxazolamine hydroxylase system is shown in Fig. 2. In this experiment, 50-g rats were given a single intraperitoneal injection of 20 mg/kg of 3,4-benzpyrene or daily injections of 75 mg/kg phenobarbital. The rats were killed at intervals, their livers were homogenized, and microsomes were assayed for ability to metabolize zoxazolamine in the presence of excess cofactors. A single injection of 3,4-benzpyrene rapidly stimulated the zoxazolamine hydroxylase system to maximum activity within 24 hr, while phenobarbital administration elevated the level of this enzyme more slowly and maximum stimulation was not observed for 3–4 days. The activity of the zoxazolamine hydroxylase system returned to control values within 2 weeks after stopping the drug. Further experiments showed that the stimulatory effect of phenobarbital or 3,4-benzpyrene on the zoxazolamine-metabolizing enzyme system correlated with the rate of zoxazolamine metabolism in the intact animal

[1] The studies reported here were carried out in collaboration with A. H. Conney and R. Welch at the Wellcome Research Laboratories, Tuckahoe, N. Y.

## TABLE I
### Environmental Chemicals which Stimulate Drug Metabolism

Drugs—phenobarbital, phenylbutazone
Carcinogenic hydrocarbons—3,4-benzpyrene
Insecticides—chlordane, DDT
Food additives—butylhydroxytoluene (BHT)
Herbicides—Herban, Diuron

(Fig. 3). Young rats received phenobarbital daily for 4 days or 3,4-benzpyrene once, 24 hr before zoxazolamine was injected intraperitoneally. Following the injection of zoxazolamine, the animals were killed at intervals. The whole skinned rat was then homogenized and the amount of remaining drug was determined. It can be seen that pretreating rats with phenobarbital, or 3,4-benzpyrene, had a dramatic stimulatory effect on the rate of zoxazolamine metabolism in the intact rat. This accelerated rate of metabolism *in vivo* correlated well with the duration

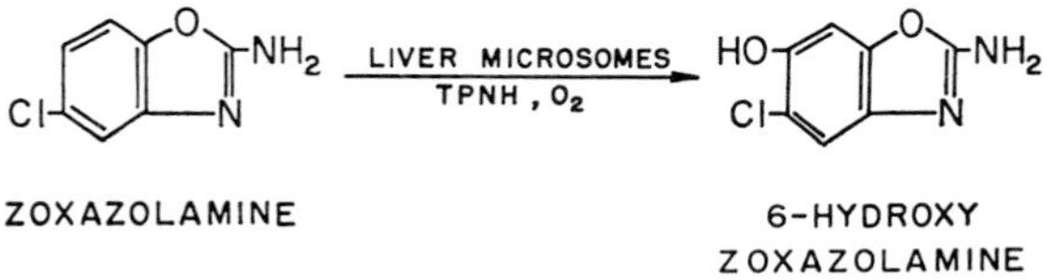

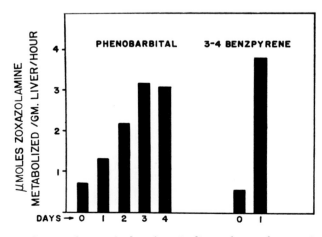

Fig. 1. Metabolism of zoxazolamine by rat liver microsomes.

of action of zoxazolamine and the activity of the microsomal enzyme system that metabolizes the drug. The duration of zoxazolamine paralysis was 730 minutes in control rats, 102 minutes in phenobarbital-treated rats, and only 17 minutes in 3,4-benzpyrene-treated rats.

Another example of the interaction of environmental agents and drugs is the ability of phenobarbital to stimulate the metabolism of the anticoagulant drugs Dicumarol and warfarin. For instance, a marked increase in the activity of the

Fig. 2. Increased zoxazolamine hydroxylase in liver of rats that received phenobarbital and 3,4-benzpyrene.

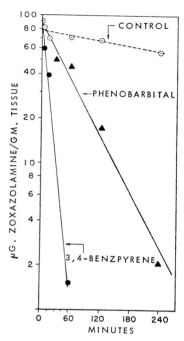

Fig. 3. *In vivo* metabolism of zoxazolamine in rats pretreated with phenobarbital or 3,4-benzpyrene.

liver microsomal enzyme which metabolizes bishydroxycoumarin (Dicumarol) occurred in rats treated with phenobarbital (8). In accord with these *in vitro* results (Fig. 4), administration of phenobarbital stimulated the rate of bishydroxycoumarin metabolism in the dog. Phenobarbital also enhanced the metabolism of bishydroxycoumarin in man, as was observed in a patient who was treated chronically with 75 mg/day of the anticoagulant (Fig. 5). When he received 1 g

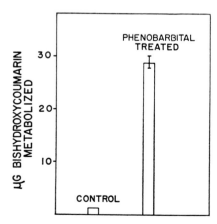

Fig. 4. Stimulatory effect of phenobarbital treatment on Dicumarol metabolism by rat liver.

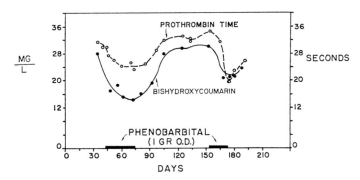

FIG. 5. Effect of phenobarbital on plasma level of Dicumarol and prothrombin response in a human subject (dose of Dicumarol, 75 mg/day).

of phenobarbital daily for 4 weeks, in addition to bishydroxycoumarin, there was a substantial lowering of the plasma level of bishydroxycoumarin and a decrease in the anticoagulant activity. Upon discontinuing phenobarbital, the plasma level of bishydroxycoumarin and the prothrombin time returned to their original values. The stimulatory effect of heptabarbital on the metabolism of some coumarins has been reported and this accounts for the inhibitory action of the barbiturate on the anticoagulant activity of the coumarins (9).

Recently, Robinson and MacDonald reported (16) that the administration of phenobarbital to patients antagonized the anticoagulant response to warfarin, which appeared to result from the ability of phenobarbital to stimulate the metabolism of warfarin. Evidence for this has now come from studies in our laboratory (12). Pretreatment of rats for 4 days with phenobarbital, DDT, or chlordane stimulates the liver enzyme which metabolizes warfarin and also enhances the drug's metabolism *in vivo* (Table II). In accord with these results, pretreatment of rats with these enzyme stimulators markedly reduces the toxicity of an acute dose of warfarin (Table III). This effect of insecticides on warfarin metabolism may explain, at least in part, the reported resistance of wild rats to the action of warfarin when used as a raticide.

TABLE II

STIMULATORY EFFECT OF PHENOBARBITAL AND INSECTICIDES ON THE HEPATIC METABOLISM OF WARFARIN *in Vitro*

Male rats weighing 50 g were injected ip twice daily for 4 days with phenobarbital or insecticide. The rats were killed on the fifth day and the 9000 $g$ supernate equivalent to 166 mg of liver was incubated with 1300 m$\mu$moles of warfarin-4-$^{14}$C for 30 minutes.

| Treatment | Daily dose (mg/kg) | Metabolite formation (m$\mu$moles) | | |
|---|---|---|---|---|
| | | 6-OH | 7-OH | 8-OH |
| Control |  | 2.7 ± 0.2 | 5.7 ± 0.5 | 1.5 ± 0.2 |
| Phenobarbital | 75 | 16.2 ± 0.6 | 29.6 ± 1.2 | 9.2 ± 0.3 |
| Chlordane | 50 | 7.2 ± 0.5 | 17.2 ± 0.9 | 4.7 ± 0.3 |
| DDT | 50 | 9.2 ± 0.5 | 16.7 ± 1.1 | 4.9 ± 0.3 |

Phenobarbital has also been shown to stimulate the metabolism of a wide variety of other drugs in experimental animals and in man. For instance, the hypnotic effect of hexobarbital and the anticonvulsant effect of diphenylhydantoin (Dilantin) are almost completely abolished in animals pretreated for several days with phenobarbital (8). In each case, phenobarbital exerts its effect by stimulating the metabolic inactivation of the subsequently administered drug. The data in Fig. 6 show the potent effect of phenobarbital in stimulating metabolism of

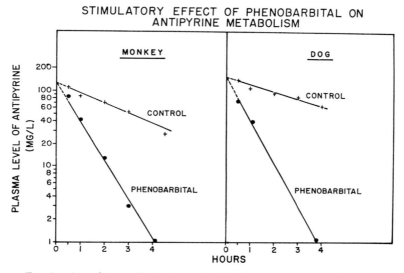

FIG. 6. Stimulatory effect of phenobarbital on antipyrine metabolism.

the analgesic drug—antipyrine—in dogs and monkeys. Treatment of rats with phenobarbital stimulated the metabolism of the antifungal drug griseofulvin (3). This effect also occurs in man: Low blood levels of griseofluvin were obtained in human subjects who received phenobarbital.

A good example of enzyme induction in man is the ability of phenylbutazone to stimulate the metabolism of aminopyrine (1). It has also been reported that treatment of humans with barbiturates accelerated the metabolism of the aminopyrine derivative, dipyrone. Patients tolerant to glutethimide metabolize the drug more rapidly than normal subjects, and the chronic administration of meprobamate in man causes an increase in the metabolism of the drug (1).

Exposure of rodents to pesticides, such as chlordane and DDT, stimulates drug-metabolizing enzyme activity and shortens the duration of action of the hypnotic drug hexobarbital (11). This effect was discovered accidentally in two different laboratories after animal quarters were sprayed with these pesticides. As mentioned earlier, pretreatment of rats with chlorinated hydrocarbon pesticides enhances the metbolism of warfarin, which results in a marked reduction in the toxicity of this anticoagulant drug (Tables II and III). Treatment of dogs with small oral doses of chlordane for 7 weeks also stimulates the metabolism of phenylbutazone (Fig. 7) and antipyrine (Fig. 8). This effect of chlordane is long-last-

TABLE III
REDUCTION OF ACUTE TOXICITY OF WARFARIN IN RATS FOLLOWING
PRETREATMENT WITH PHENOBARBITAL AND PESTICIDES

| Pretreatment | Daily dose[a] (mg/kg) | $LD_{50}$ (mg/kg) |
|---|---|---|
| None | — | <30 |
| Phenobarbital | 75 | >300 |
| Chlordane | 50 | >300 |
| DDT | 50 | >300 |

[a] Compounds given ip to 50-g male rats for 4 days.

ing; dogs metabolize these drugs at an accelerated rate for a considerable period after chlordane administration has been discontinued, presumably because of the retention of the pesticide in the animal's body fat. Although these observations, which show the potent effect of pesticides on drug metabolism, are important for animal experiments, their relevance to drug therapy in man remains to be determined.

Recent studies showed that drugs may also interact in the metabolism of a variety of steroids. Evidence has been found that steroid hormones are normal body substrates for oxidative drug-metabolizing enzymes in liver microsomes (14) and, consequently, drugs that stimulate the microsomal oxidation of drugs also stimulate the microsomal hydroxylation of steroids (6). Thus, treatment of rats with phenobarbital, chlorcyclizine, phenylbutazone, and other commonly used drugs stimulated several-fold the liver microsomal hydroxylation of testosterone, estradiol-17$\beta$, progesterone, and desoxycorticosterone. Recently, it was found that treatment of human subjects with phenobarbital or diphenylhydantoin markedly

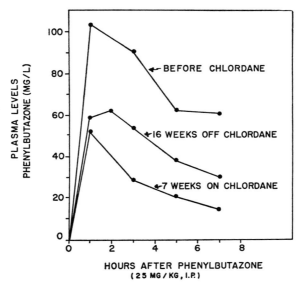

FIG. 7. Stimulatory effect of chlordane (5 mg/kg, po, three times weekly) on phenylbutazone metabolism in the dog.

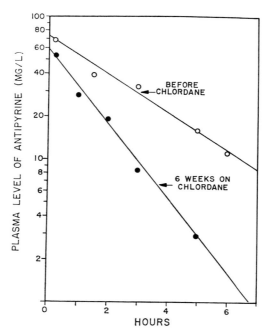

Fig. 8. Stimulatory effect of chlordane (5 mg/kg po, three times weekly) on antipyrine metabolism in the dog.

increased the extra-adrenal metabolism of cortisol to 6-hydroxycortisol, which was excreted in the urine. An explanation for the stimulatory effect of these drugs on cortisol metabolism in man comes from the finding that treatment of guinea pigs with diphenylhydantoin or phenobarbital for several days stimulated the formation of enzymes in liver microsomes that hydroxylate ($6\beta$) cortisol. The physiologic importance of drug-induced stimulation of steroid hydroxylation in liver must await further investigation.

Treatment of rats with drugs, halogenated hydrocarbon insecticides, and other foreign chemicals not only stimulates the activity of drug- and steroid-metabolizing enzymes in liver microsomes, but also elicits several other biochemical responses (4). These include enhanced liver growth and synthesis of liver microsomal protein, proliferation of hepatic smooth-surface endoplasmic reticulum, and enhanced metabolism of glucose via the glucuronic acid pathway to ascorbic acid. The significance of these various biochemical responses elicited by the administration of environmental chemicals remains to be established (7). However, it should be noted that the stimulatory effects of drugs and insecticides on the drug metabolism in experimental animals is paralleled by an altered duration and intensity of drug action, since the stimulatory effect of drugs on drug and steroid metabolism occurs not only in animals, but also in man. The effect of insecticides on these biochemical responses requires careful evaluation to determine whether the environmental exposure of man to insecticides can alter his response to drugs or steroid hormones.

## REFERENCES

1. BURNS, J. J. (1964). Implications of enzyme induction to drug therapy (Editorial). *Am. J. Med.* **37**, 327–331.
2. BURNS, J. J., CUCINELL, S. A., KOSTER, R., AND CONNEY, A. H. (1965). Applications of drug metabolism to drug toxicity studies. *Ann. N. Y. Acad. Sci.* **123**, 273–286.
3. BUSFIELD, D., CHILD, K. J., ATKINSON, R. M., AND TOMWICH, E. G. (1963). An effect of Phenobarbitone on blood levels of griseofulvin in man. *Lancet* **2**, 1042–1043.
4. CONNEY, A. H., AND BURNS, J. J. (1963). Induced synthesis of oxidative enzymes in liver microsomes by polycyclic hydrocarbons and drugs. *In* "Symposium on Regulation of Enzyme Activity and Synthesis in Normal and Neoplastic Liver" (G. Weber, Ed.), pp. 189–214. (Macmillan), New York.
5. CONNEY, A. H., DAVISON, C., GASTEL, R., AND BURNS, J. J. (1960). Adaptive increases in drug-metabolizing enzymes induced by phenobarbital and other drugs. *J. Pharmacol. Exptl. Therap.* **130**, 1–8.
6. CONNEY, A. H., SCHNEIDMAN, K., JACOBSON, M., AND KUNTZMAN, R. (1965). Drug induced changes in steroid metabolism. *Ann. N. Y. Acad. Sci.* **123**, 98–109.
7. CONNEY, A. H., WELCH, R. M., KUNTZMAN, R., AND BURNS, J. J. (1967). Pesticide effects on drug and steroid metabolism: A Review. *Clin. Pharmacol. Therap.* **8**, 2–10.
8. CUCINELL, S. A., CONNEY, A. H., SANSUR, M., AND BURNS, J. J. (1965). Drug interaction in man: 1. Lowering effect of phenobarbital on plasma levels of bishydroxycoumarin (Dicumarol) and diphenylhydantoin (Dilantin). *Clin. Pharmacol. Therap.* **6**, 420–429.
9. DAYTON, P. G., TARCAN, Y., CHENKIN, T., AND WEINER, M. (1961). The influence of barbiturates on coumarin plasma levels and prothrombin response. *J. Clin. Invest.* **40**, 1797–1802.
10. GOLBERG, L. (1966). Liver enlargement produced by drugs: Its Significance. *Proc. European Soc. Study Drug Toxicity (Rome)* **7**, 171–184.
11. HART, L. G., SHULTICE, R. W., AND FOUTS, J. R. (1963). Stimulatory effects of chlordane on hepatic microsomal drug metabolism in the rat. *Toxic Appl. Pharmacol.* **5**, 371–386.
12. IKEDA, M., CONNEY, A. H., AND BURNS, J. J. (1968). Stimulatory effect of phenobarbital and insecticides on warfarin metabolism in the rat. *J. Pharmacol. Exptl. Therap.* **162**, 338–343.
13. KINOSHITA, F. K., FRAWLEY, J. P., AND DUBOIS, K. P. (1966). Effects of subacute administration of some pesticides on microsomal enzyme systems. *Toxic Appl. Pharmacol.* **8**, 345–346.
14. KUNTZMAN, R., JACOBSON, M., SCHNEIDMAN, K., AND CONNEY, A. H. (1964). Similarities between oxidative drug-metabolizing enzymes and steroid hydroxylases in liver microsomes. *J. Pharmacol. Exptl. Therap.* **146**, 280–285.
15. REMMER, H. (1962). Drugs as activators of drug enzymes. *Proc. First Intern. Pharmacol. Meeting (Stockholm)* **6**, 235–246. Macmillan, New York.
16. ROBINSON, D. S., AND MACDONALD, M. G. (1966). The effect of phenobarbital administration on the control of coagulation achieved during warfarin therapy in man. *J. Pharmacol. Exptl. Therap.* **153**, 250–253.

# Difficulties in Extrapolating the Results of Toxicity Studies in Laboratory Animals to Man

## David P. Rall

*National Cancer Institute, National Institute of Health, Bethesda, Maryland*

I am pleased that the title of this presentation demands that I only discuss the difficulties in extrapolating the results of animal studies to man, and not that I propose solutions to these problems. This is a difficult topic to consider in a scientific and rigorous fashion and, therefore, generally is treated in an anecdotal fashion. Any pharmacologist or toxicologist can (and will) remember, discuss, and clearly document individual instances of either a perfect or perfectly imperfect correlation of results between laboratory animals and man. I shall attempt to deal more broadly with this topic and analyze these difficulties by covering three different but related areas. The first concerns the pharmacologic basis for prediction; the second, how good the prediction is; and the third, how good the prediction can and should be. This last may be paraphrased: What are we really worried about?

In considering extrapolation of biologic data from one species to another, it is worthwhile to consider the definition of a species. A species is a group of individuals with a number of common characteristics different from other groups of individuals. Variations around these common characteristics should not be in large quantum jumps, but continuously variable. Differences due to sex and age, of course, are present, and those due to polymorphism also occur. One key characteristic of a species is that the individual can interbreed. Thus, each species, by definition, has some characteristics of uniqueness. It would be surprising, indeed, if some aspects of this uniqueness were not related to responses to chemical agents.

Pharmacologists are beginning to recognize, if not understand, some of the factors which permit simple and successful extrapolation of some aspects of toxicity data from one species to another. It is, however, important to point out at the outset that there are many potential differences which are not related to differences between species, such as age, polymorphism, size, diet, environment, schedule, and route of chemical administration, state of health, and extent of supportive treatment. Consider age. It is common for toxicity studies to be performed on weanling animals. When the results with weanling animals fail to predict for either very young or very old humans, interspecies differences are often involved. Although I shall not document each instance, it is clear that many differences between the human, either well or sick, and the experimental animal are not related to species differences. I would urge anyone considering this topic to pay particular attention to these factors.

There are differences and similarities known to be due to physiologic, biochemical, and anatomic differences among species. Absorption, distribution and storage, excretion, metabolism, and site and mechanism of action are all involved in the pharmacologic action of a chemical.

It is informative to analyze these five steps in relation to the degree of consistency among various species. For instance, the rate of absorption of an agent after intraperitoneal administration in mice and rats is known to be similar to that after intravenous administration in man. Intraperitoneal absorption for elasmobranch fish, however, is slow and erratic (9). In general, absorption of drugs is comparable among vertebrate species. The distribution and storage of compounds once they have been absorbed also tend to be comparable in a variety of vertebrate species. Some differences are known with relation to plasma protein binding. Dr. Schanker has considered these areas in detail. In terms of excretion, there are few differences between the common laboratory animals and man. It is, however, well known that some marine teleosts are without glomeruli, and recently it has been shown that the cyclostome, *Myxine glutinosa*, may be considered to be atubular with respect to organic acid excretion (12). The metabolism of drugs is, however, far from comparable from species to species, as Dr. Williams and Dr. Burns documented. Not only are different metabolites formed, but, when the same metabolites are formed, they may form at different rates from one species to another. It is in this area that interspecies comparisons break down. The mechanism or site of action of compounds is fairly comparable from species to species (1).

Thus, with the important exception of drug metabolism, there are few known differences between species that might influence the predictability of toxicity data. I have considered only gross differences, yet I am certain that there are subtle as-yet-unelucidated differences between common laboratory animals and man. As we discover these, not only will we be able to extrapolate with more precision from one species to another, but, as has been fruitful in the past, the elucidation of the mechanism of these differences will allow the definition of basic biochemical and physiologic mechanisms.

There are, therefore, five basic steps in the ultimate action of the compound. Let us consider that, for each of the five steps—absorption, distribution, excretion, metabolism, and mechanism of action—the correlation coefficient between the pharmacology of a compound in two different species is positive and 0.90, a very high value. $0.9^5$ should then estimate the over-all correlation in drug response between the two species. This is about 0.58. Although positive, it is hardly good. In general terms, this would suggest that in 58% of the instances there would be perfect correlation. Alternatively, let us consider that for metabolism there is only a correlation coefficient of 0.5, but that for the other steps it is 0.95. In this instance, the overall would be about 0.42, and this is not good predictability.

There are really two parts to the question of how good the prediction is. The first is the quantitative, the second, qualitative. Quantitative predictability is concerned with the consistency of the therapeutic or toxic doses between species, and qualitative, with the nature of the pharmacologic response. Let us first consider some aspects of quantitative predictability. To explore this, we studied a number of antineoplastic drugs with respect to their tolerated dose in man, monkey, dog, rat, and mouse (4). One interesting aspect of this comparison is that these agents are generally not involved in the variable drug-metabolizing systems and the variability in metabolism is very much less between species. In a real sense, this study tested variation as a function of size and of absorption, distribution, excretion, and mechanism of action. The agents used were alkylating agents, antimetabolites, and antibiotics (listed in Table I). Figures 1 and 2 demonstrate the correlation between one strain of mouse and man and between the rhesus monkey and man. Over a three-fold log range of absolute dosage, the correspondence was very good. The correlation coefficients ranged well above 0.9.

TABLE I
Drugs Used

Actinomycin D: NSC-3053
Alanine mustard: NSC-17663; dl-alanine, $N,N$-bis(2-chloroethyl)-,hydrochloride.
Amethopterin: NSC-740; glutamic acid, $N$-[$p$-[[(2,4-diamino-6-pteridinyl)methyl]methylamino]-benzoyl]-.
BCNU: NSC-409962; urea, 1,3-bis(2-chloroethyl)-1-nitroso-.
Cytoxan: NSC-26271; 2H-1,3,2-oxazaphosphorine, 2-[bis(2-chloroethyl)amino]tetrahydro-, 2-oxide, hydrate.
5-Fluorouracil: NSC-19893.
5-FUDR: NSC-27640; uridine, 2′-deoxy-5-fluoro-.
Hydroxyurea: NSC-32065.
6-Mercaptopurine: NSC-755; purine-6-thiol, hydrate.
Methyl-GAG: NSC-32946; guanidine, 1,1′-[(methylethanediylidine)dinitrilo]di-, dihydrochloride, hydrate.
Mitomycin C: NSC-26980; carbamic acid, ester with 6-amino-1,1a,2,8,8a,8b-hexahydro-8-(hydroxymethyl)-8a-methoxy-5-methylazirino[2′,3′:3,4]pyrrolo[1,2-$a$]-indole-4,7-dione.
Myleran: NSC-750; 1,4-butanediol, dimethanesulfonate.
Nitrogen mustard (HN2): NSC-762; diethylamine, 2,2′-dichloro-$N$-methyl-, hydrochloride.
Nitromin: NSC-10107; diethylamine, 2,2′-dichloro-$N$-methyl, $N$-oxide, compd. with hydrochloride (1:1).
l-Phenylalanine mustard: NSC-8806; l-alanine, 3-[$p$-[bis(2-chloroethyl)amino]phenyl]-, hydrochloride.
ThioTEPA: NSC-6396; phosphine sulfide, Tris(1-aziridinyl)-.
Vinblastine: NSC-49842; vincaleukoblastine, sulfate, hydrate.
Vincristine: NSC-67574; leurocristine, sulfate.

Figure 3 shows the same data relating mouse to man as expressed in milligrams/kilogram. The relationship on a milligram/kilogram basis is just as good as on a milligram/square-meter basis. The difference lies in the location of the cluster of points. It is moved to the right by a 12-fold factor and the line of best fit does not go through the origin. A similar 12-fold factor relates surface area to body weight between mouse and man. For at least one drug (methotrexate), it can be shown that the integrated plasma concentration time function after the same milligram/kilogram dose is about 10 times higher in man than in mouse. The results underscore the need to study the metabolism and plasma concentration of chemicals in laboratory animals and in man.

Qualitative aspects of predictability for anticancer drugs also have been

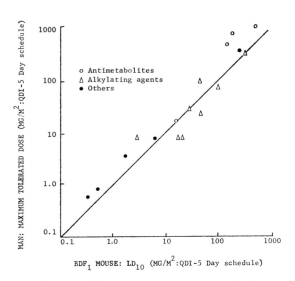

Fig. 1. Comparison of toxicity data on anticancer agents for the mouse and man (on a $MG/M^2$ basis). (With acknowledgements to Cancer Chemotherapy Reports.)

studied. The predictability was quite good for such systems as bone marrow, liver, kidney, and lung (10). Many, but not all, nervous-system effects which were present in man were missed in animals, and in this area the predictability was only fair. Animals rarely predicted important dermatologic toxicity in man. Six therapeutic agents (not anticancer drugs) have been studied for the qualitative predictability from animals to man (8). In general, there was surprisingly good agreement from animals. It is important to note that the animals failed in

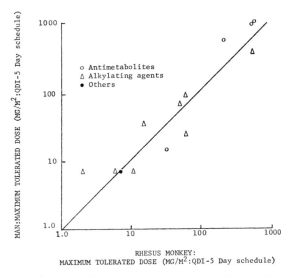

Fig. 2. Comparison of toxicity data on anticancer agents for the rhesus monkey and man (on a $MG/M^2$ basis). (With acknowledgements to Cancer Chemotherapy Reports.)

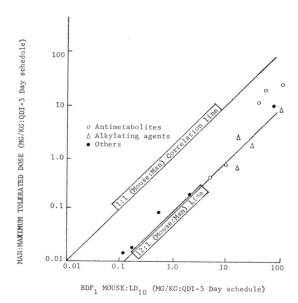

Fig. 3. Comparison of toxicity data on anticancer agents for the mouse and man (on a MG/KG basis). (With acknowledgements to Cancer Chemotherapy Reports). The 12:1 relationship shown on a MG/KG basis is equivalent to the 1:1 relationship shown on a MG/M² basis (Chart 4). The approximate 12:1 relationship (mouse:man) is in agreement with the ratio of the KM factors used for these species, i.e., 37:3 (man:mouse) = ca. 12.

two major areas. Rats and dogs could not predict some symptoms uniquely human; headache, loss of libido, etc., are difficult to elicit from laboratory animals. Furthermore, animals are given deliberately lethal doses of the drug and can show signs, or pathologic conditions, that are not seen in man. It is not, to my way of thinking, a serious problem if the laboratory animals exhibit pathology which ultimately is not seen in man. It is more important to consider the serious pathologic conditions seen only in man, which would not be anticipated from the animal study. Fortunately, in this very small series, this happened rarely.

A significant point in comparing laboratory animals and human data is, I am convinced, the undeniable fact that the clinical acumen of a skilled physician caring for an ill patient surpasses that of the most careful toxicologist studying a variety of animals in a toxicology study. I would like to illustrate this by reference to the common statement issued whenever the topic of interspecies comparisons is considered. Someone noted that penicillin is so toxic in the guinea pig that, had it first been tested in this animal, it might have been discarded (7). The facts are interesting, as always. First, it is clear that, while the single-dose toxicity of penicillin is about the same for guinea pigs as for other animals, the apparent differences or discrepancies occur in repeated-dose or subacute toxicity studies (5). At relatively low repeated dosage, significant mortality occurred in guinea pigs. This mortality was usually 50–80% and rarely, if ever, 100%. The guinea pigs became tolerant to this, and newborn or very young guinea pigs (3) do not show this apparent susceptibility. A careful examination of the mechanism of this mortality has been underway for a number of years and its seems to involve, in some way, superinfection or overgrowth of bacteria in the guinea pig. Thus, it is not likely to be a direct toxic effect of penicillin, but a secondary effect on the bacterial flora of the guinea pig. Since leptospirosis, an otherwise lethal infection, can be cured by the use of penicillin in the guinea pig, the toxicity is not so prohibitive that penicillin cannot be used as a therapeutic agent in that species (6). More important, the phenomenon of the bacterial superinfection is seen in man and is one of the major clinical problems associated with the use of penicillin in man (13). The toxicity of penicillin in the guinea pigs did, in fact, predict very well a serious toxic effect of penicillin in man. However, I am more concerned about the lack of ability of the pharmacologists to appreciate the prediction.

Let us now consider briefly how good our predicting systems can or should be. Let us assume that 100 new compounds are introduced into man every year. With one of these 100 compounds, there is a failure of prediction in which either death or severe disability occurs. This is considered by the lay press to be a catastrophe. Yet, a predictability score of 99% should be considered very good. The problem of predictability lies primarily in the fact that the stakes are so high and not that the systems are so bad.

Let us explore where the major hazards lie in the drug development process. I would like to suggest that the toxicity in the initial exploration in man of a new drug is not likely to be a public-health problem. Few patients are used, and they are under tightly controlled clinical situations. While this is true with the use of therapeutic or diagnostic drugs, it is not true with pesticides and other agricultural chemicals.

For pesticides and other similar agricultural chemicals, there is no initial clinical trial, intentionally at least. Then how does one attempt to relate toxicity from animals to man? If something is known of the plasma concentrations and metabolic fates of these agents, estimates can be made on the basis of comparable plasma concentrations. If metabolism of the chemical is unknown or if it is not metabolized to any significant extent, the relationship will, on the average, tend to follow some fractional power of the body weight similar to the surface-area relationship.

The major problems lie in phases in which large numbers of patients are exposed to the agents, drugs, or agricultural compounds, often under conditions of less than perfect patient control. The nature of the toxicity is worth considering in this context. What sort of toxicity are we worried about? I would suggest there are three instances in which the toxicity can and will be serious and a potential public-health problem. The first is that of irreversible toxicity. Many toxicities from drugs, whether potentially serious or not, are reversible if the agent is stopped in time. With simultaneous exposure to a multitude of therapeutic, diagnostic, and environmental agents, it may be difficult to determine which is the toxic agent and should be stopped. One cannot equate a mild reversible toxic effect, which can easily be ameliorated with the cessation of the causative agent, with a serious public-health problem. In this instance, once the toxicity is noted, it will proceed to its inevitable lethal or permanently damaging manifestation.

Second, I am concerned about toxicity which occurs only rarely, i.e., has a relatively low incidence. This is particularly dangerous because it can rarely be predicted by animal testing. It must be remembered that only about 1000 animals are used in the toxicity evaluation of any drug, and yet that same drug may be administered to many millions of persons. It should be added clearly that an increase in number of animals is not the proper answer. This may be toxicity only to one segment of population. With thalidomide, for instance, only women pregnant a certain number of days were susceptible. Alternatively, this may be related to genetic traits or it may be related to the concomitant and possibly unusual use of two drugs. There are many instances in which joint toxicity is serious, but the toxicity of either drug alone is not.

The third instance which worries me is that in which there is a great delay between the administration of the causative agent and the manifestation of the toxic effect. In this case, it is difficult to relate cause and effect, and the agent may not be detected until much damage is done.

Careful studies with laboratory animals can and usually will predict the possibility of irreversible toxicity. Pharmacologists, environmental physiologists, and clinicians must be aware, however, of the hazards of irreversible toxicity. Similarly, careful studies with laboratory animals can and usually will produce delayed toxicity. Laboratory animals, however, are unlikely to aid in a clinically useful way in the prediction of low-incidence toxicities. Let us hope that pharmacologists, environmental physiologists, and clinicians will cogitate earnestly over all new agents so that they might predict peculiar toxicities that might develop. Equally important, I am convinced, is the urgent need for effective implementation of a well-devised scheme for monitoring clinical drug use in the general population (11, 2). We shall always attempt to extrapolate animal toxicological data successfully to man and prevent the unnecessary exposure of man to the deleterious effects of chemicals. No biological process is perfect, therefore we must have an effective monitoring system for chemical or drug toxicity at the clinical level.

## REFERENCES

1. BRODIE, B. B., COSMIDES, G. J., AND RALL, D. P. (1965). Toxicology and the biomedical sciences. *Science* **148**, 1547.
2. FINNEY, D. J. (1965). The design and logic of a monitor of drug use. *J. Chronic Diseases* **18**, 77.
3. FOLLETT, D. A., AND BATTISTO, J. R. (1963). Alteration of guinea pig response to lethal doses of penicillin by pretreatment of test animals. *Bacteriol. Proc.* p. 70.
4. FREIREICH, E. J., GEHAN, E. A., RALL, D. P., SCHMIDT, L. H., AND SKIPPER, H. E. (1966). Quantitative comparison of toxicity of anticancer agents in mouse, rat, hamster, dog, monkey, and man. *Cancer Chemotherapy Rep.* **50**, 219.
5. HAMRE, D. M., RAKE, G., MCKEE, C. M., AND MACPHILLAMY, H. B. (1943). The toxicity of penicillin as prepared for clinical use. *Am. J. Med. Sci.* **206**, 642.
6. HEILMAN, F. R., AND HERRELL, W. E. (1944). Penicillin in the treatment of experimental leptospirosis icterohaemorrhagic (Weil's disease). *Proc. Mayo Clinic* **19**, 89.

7. Koppanyi, T., and Avery, M. A. (1966). Species differences and the clinical trial of new drugs: A review. *Clin. Pharmacol. Therap.* **7,** 250.
8. Litchfield, J. T. (1961). Forecasting drug effects in man from studies in laboratory animals. *J. Am. Med. Assoc.* **177,** 104.
9. Litchfield, J. T. (1939). The effects of sulfanilamide on the lower vertebrates. *J. Pharmacol.* **67,** 212.
10. Owens, A. H. (1963). Predicting anticancer drug effects in man from laboratory animal studies. *J. Chronic Diseases* **15,** 223.
11. Rall, D. P. (1964). An early warning system designed to detect hazards of drug usage. *Toxicol. Appl. Pharmacol.* **6,** 366.
12. Rall, C. P., and Burger, J. W. (1967). Some aspects of hepatic and renal excretion in myxine. *Am. J. Physiol.* **212,** 354.
13. Weinstein, L. (1964). Superinfection: A complication of antimicrobial therapy and prophylaxis. *Am. J. Surg.* **107,** 704.

# Some Prospects in Toxicology

BERNARD B. BRODIE

*National Heart and Lung Institutes,
National Institutes of Health, Bethesda, Maryland*

My remarks, which also represent the thoughts of Dr. Horvath, concern some future directions in research on environmental chemicals. Such agents are dangerous if they produce biochemical or structural lesions in a significant number of people. In the past, toxicology—whether it concerns environmental chemicals or those used as medicine—has suffered from expediency. It has been under constant pressure to solve particular problems and as a result has not been encouraged to develop general principles. A major problem facing toxicologists, as pointed out by Rall, Burns, and others, is: How can the risk of an environmental agent be assessed in animals? Meaningful extrapolation of information from animal studies to man can only come from a basic understanding of the mechanisms whereby chemicals interact with biochemical processes in the body. Furthermore, little progress can be made in studies of toxicity mechanisms, unless the biological response can be correlated with the actual amount of the toxic agent in the body. Toxicology, as well as pharmacology, is often criticized for not being a more quantitative science. All too often, toxic effects are reported without regard for the concentration of substance that produces these effects. But pharmacology and toxicology are young disciplines. It is only in relatively recent times that pharmacologists could work with pure substances—rather than with plant, animal, and mineral extracts—and could relate drug effects to physiologic characteristics. In the study of environmental chemicals, this problem persists since we often deal with intricate and inconstant mixtures, like smog, smoke, or inhalants from cigarettes, or with substances whose chemical identity is known but which produce their toxic effects through active metabolites. Thus, the nature of the toxicant is often unknown; we do not even know what to study.

Another apparent obstacle arose when pharmacologists, studying pure drugs which had finally become available, found to their amazement that there are tremendous species differences in intensity of responses. Curiously, these were first ascribed to differences in the sensitivity of receptor sites; little thought was given to the possibility that at least part of the species variation might be attributed to differences in the way a substance was handled by the animal and the resulting differences in the amount of a substance available to the receptor. Now we are aware of the wide species variability in the activities of the drug-metabolizing enzymes and this has become an important factor in the precise study of mechanisms.

This new emphasis on precision is vital. If anyone doubts the importance of quantitation to toxicology, he should reflect on what biochemistry would be like today if, for example, the complicated interaction of lipid and carbohydrate metabolism was studied without methods for measuring free fatty acids and glucose, for determining the fate of those substrates, and for measuring the effects of inhibitors on enzymes that are responsible for their synthesis or disappearance. Great advances in a biologic discipline have usually been associated with the introduction of precise techniques. In pharmacology and toxicology this attitude is not yet sufficiently prevalent. With environmental substances, as I have noted, the problem is even more difficult because in many cases the identities of the chemical producing the damage is not known. Here is an opportunity for gifted chemists to engage in the isolation and identification of trace substances in materials like smog and tobacco smoke, as well as possible toxic derivatives of these substances formed in the body. Over the last few years, instrumentation has developed that staggers the imagination: countercurrent distribution, thin-layer chromatography, gas chromatography, mass spectrometry, and many other techniques which permit the isolation and identification of organic compounds in microgram amounts. Another encouraging advance has been the development of methods for the assay of small amounts of substances in biologic fluids and tissues. Simple methods are now becoming available for the quantitative assay of nanograms and, in some cases, picograms of material. The use of isotope derivatives is such a method. By this procedure one can, for example, extract an unlabeled primary or secondary amine present in body fluids into a solvent, and then acetylate it with highly labeled acetic anhydride. Specificity is not usually a major problem with substances that are extractable into nonpolar solvents, and the sensitivity of this procedure may be quite high—hundreds of counts per nanogram. Recently, an exciting new technique has appeared that does not depend on radioactivity. Surprisingly enough, this technique has been used for a number of years in the assay of chlorinated insecticides by agriculturalists, but owing to communication difficulties between disciplines, it took a long time for pharmacologists to learn about it. It is a gas chromatographic method, specific for halide or nitro groups, involving the use of an "electron-capture" detector. Extremely high sensitivity with almost no blank is obtained. There are simple ways of introducing a halide or nitro group into a compound if it does not already contain one. Thus, one can foresee the possibility of an almost "universal" method, at least for compounds that are lipid-soluble, and perhaps for other substances as well. Those working with insecticides are already using this technique to assay picogram amounts.

What is our progress to date in the elucidation of the biochemical mechanisms of action of environmental hazards? We know, of course, the mechanism of action of such substances as cyanide and carbon monoxide. In addition, we know that some hydrazines, such as isoniazid, which is used in the treatment of tuberculosis, produce their toxicity by reacting with pyridoxal phosphate. But it is proving extremely difficult to get at the mechanism of a substance like carbon tetrachloride. Lately, however, a note of optimism has been sounded. Recent work is providing clues which point to future directions for research. People have become aware that many substances which produce toxic manifestations do not act as such but are activated in the body to toxic substances. In other words, the chemical pathways in the body, which we formerly called "detoxication" processes, serve to convert foreign substances to less lipid-soluble substances which can be excreted; but sometimes these metabolites are much more toxic than the original substance. One illustration is N-hydroxylation brought about by the action of a liver microsomal enzyme which puts an OH group directly onto nitrogen; for example, acetanilide forms phenylhydroxylamine, a potentially toxic substance. The antimalarial agents—primaquine and pamaquine—are converted by N-hydroxylation to extremely potent oxidizing agents which can also lyse red blood cells. Hydroxylamines have been implicated in hemolytic anemia, methemoglobinemia, carcinogenesis, and even in allergic phenomena.

The implication of hydroxylamines in hemolytic anemia illustrates another factor in the elucidation of mechanisms of toxicity—the genetic disposition of some persons to agents which may not be toxic to others. We are aware that a number of aromatic amines precipitate acute hemolysis in persons with an inherited deficiency of glucose-6-phosphate dehydrogenase in their red cells. This enzyme is responsible for the production of TPNH, a substance needed by the cell to maintain glutathione in a reduced state. Without TPNH, the whole cell falls apart. In the normal person these hydroxylamines or quinones cause no harm because there is no shortage of TPNH in the red cell. With the genetic deficiency, there is only a small amount of TPNH, and since the toxic metabolites compete for this substrate, the cells disrupt. Thus, it is the genetic deficiency together with trace amounts of toxic metabolites that produces the trouble. This raises the possibility that other environmental agents elicit toxic effects in some persons by interacting in the body when there is a deficiency of substrate production. Normally, this deficiency causes no harm, but when a substance foreign to the body competes for the use of this substance, then a lesion can occur, simply because the metabolic machinery in a particular organ is impaired.

There is another genetic lesion, a very strange one, which accounts for porphyria. Persons who have latent porphyria have a remarkable biochemical anomaly: The body has the capacity to overinduce ALA synthetase—the enzyme that catalyses the rate-limiting step in the formation of porphyrins. When these persons are given certain drugs, ALA synthetase goes wild and more and more enzyme is produced. The biochemical machinery for making porphyrins gets clogged by an excess of $\delta$-aminolevulinic acid. As a result, abnormal porphyrins are formed and cause photosensitive reactions. The examples of porphyria and hemolytic anemia emphasize the importance of having biochemists who are also interested in genetics look for enzyme deficiencies in particular organ systems whose functions collapse under the influences of certain environmental hazards, but will do so only in certain predisposed people.

Since hydroxylamines, as noted above, may be potent carcinogens, the N-hydroxylating enzyme is now being invoked by cancer researchers to explain why certain substances are highly carcinogenic. For example, 2-acetylaminofluorane is converted to the N-hydroxyl derivative, thought to be the actual carcinogen. Of particular interest is the fact that the N-hydroxylation of naphthylamine, a substance quite close in structure to aniline, results in a potent carcinogenic substance.

A particular source of concern in disclosing potential carcinogens are compounds with hidden alkylating or acylating properties—one would never suspect them as carcinogens. For example, the naturally occurring carcinogenic pyrrolizidine alkaloids, present in certain nuts and eaten by natives in Africa, are esters. One would not ordinarily think of an ester as an alkylating agent. Yet the chemical configuration is such that they can alkylate SH groups. Even a substance as simple as ethylcarbamate (urethane) has been implicated as an alkylating agent through its conversion to an $N$-hydroxylated product. Who would have guessed that dimethylnitrosamine would be an alkylating agent? Yet Dr. Magee has shown that in the body this compound is demethylated; the resulting product is a very efficient alkylating agent with a structural resemblance to diazomethane, long used by organic chemists as a potent methylating agent.

Another difficult problem is the fact that alkylating or acylating agents produce a wide spectrum of responses. For example, the adrenergic blocking agent, dibenzyline, is an alkylating agent. If it were a new drug, we would hesitate to give it to patients because we are aware of the toxicity of alkylating agents; nevertheless, it seems to be a relatively safe agent and acts rather specifically at adrenergic receptors. Thalidomide, in contrast, is an acylating agent which exerts its effect mainly on the fetus. On the other hand, many acylating agents can be antigrowth or antitumor agents. Thus, alkylating and acylating agents can be teratogenic, carcinogenic, mutagenic, or even antitumor agents. What is the relationship? Why do these substances yield such varied responses? These questions are important to the study of environmental hazards because of the nature of compounds found in smog, gasoline fumes, etc.

Future research on mechanisms of toxicity will lead to further understanding of hypersensitivity or allergic responses. Since allergic responses to drugs are so widespread, one must assume that there is hypersensitivity to many chemicals in the environment. Allergic states involve the formation of an antigen through the stable union of a chemical or its metabolite with a protein. Parker and his group at Washington University in St. Louis have demonstrated that penicillin, which has an unstable lactam ring, can combine nonenzymatically with protein to form an antigen. It has also been postulated that hydroxylamines may be involved in the formation of antigens; this might explain the antigen-antibody reactions to so many primary and secondary amines.

However, blaming every toxic effect that we don't understand on hypersensitivity is not a very helpful approach. All too often, when an antigen-antibody reaction cannot be established, or when an allergic response cannot be obtained on rechallenging the subject with a trace of drug, the attitude of many workers in this field is to rationalize some reason for failure of the antigen-antibody reaction, rather than to accept the possibility that it was not an allergic response in the first place. For example, 10 years ago hemolytic anemia induced by a drug was arbitrarily concluded to be an allergic response. Recently, a pharmaceutical house developed a very active antirheumatic agent, the most potent it had ever had. It was given to patients and in 7 days 30% of the patients had a skin eruption. Was one to call this a hypersensitivity reaction when it developed in 30% of the patients within 1 week, The possibility must be entertained that some substance, or perhaps a biotransformation product, acted directly to produce a lesion.

Finally, there is the question of drug interactions with environmental conditions.

These are important because patients are being treated with so many potent drugs. Dr. Westermann mentioned the hit-and-run drugs such as the monoamine oxidase (MAO) inhibitors—drugs that inactivate an enzyme and disappear from the body within a short time. A patient who has been given an MAO inhibitor may switch doctors. If the second doctor doesn't know that the patient had received an MAO inhibitor, he may give one of a number of drugs that are potentially dangerous to this patient. Especially dangerous would be imipramine, which is another antidepressant drug. This drug by itself is relatively nontoxic but can be very dangerous in a patient in whom MAO has been blocked. I am surprised that some doctors still play Russian roulette and give these two drugs together. Many drugs affect the sympathetic system by depleting the catecholamines. Immediately, one realizes that these drugs might be inimical to a patient placed in a situation in which he requires a reflex reaction utilizing the sympathetic system. For example, if an animal is deprived of sympathetic function by means of drugs and is given another drug that acts on the microcirculation, the blood pressure may fall drastically and the animal may die. Sympathetic reflex mechanisms fail, because there is no norepinephrine to mediate them. Similarly, patients, especially elderly ones who had been given reserpine, did not react very well that cold winter in England a number of years ago because the reserpine had depleted the body stores of ACTH and catecholamines.

In conclusion, our environment is becoming contaminated by a profusion of substances in the form of industrial and municipal wastes, air and water pollutants, herbicides, pesticides, cosmetics, food additives, as well as drugs administered over extended periods of time whose possible poisonous effects are difficult to predict from animal data. Until now each problem has been tackled more or less independently. Lacking unified and simplified generalizations, attempts have been made to solve each problem by applying toxicity tests, often inadequate, to large masses of animals for long periods of time. Meanwhile, the number of chemicals that affect man increases at an alarming rate. It has become essential, therefore, to disclose the basic mechanisms by which these substances interact with biological systems. It is hoped that chemists, molecular biologists, physiologists, and pharmacologists will work together to provide a suitable framework by which cause and effect mechanisms can be understood.

# Effects of Environmental Agents at the Genome Level[1]

R. K. BOUTWELL

*McArdle Laboratory, University of Wisconsin Medical School, Madison, Wisconsin 53706*

Evidence is reviewed on the components of the carcinogenic process, as a frequent specific effect of environmental agents. As early as 1944 it was shown that the carcinogenic process consists of two steps: (1) initiation of an irreversible change, presumably in the genome of the cells exposed; and (2) promotion of tumor formation made possible by that change. Detailed studies revealed that in addition to having irreversible effects, the initiator given in fractional doses is additive. The converse is true of the promoter—tumors fail to appear if the interval between successive fractional doses is lengthened. These observations suggest that the two processes are qualitatively different in their action, that each has a different molecular target within the cell, and that each has its unique contribution to carcinogenesis. Many substances seem to incorporate both components, but others exhibit the properties of initiator or promoter separately.

The irreversible, heritable (within the cell line), and dose-dependent nature of the initiator effect suggested that it brought about a change in the coding properties of DNA. Experiments with the agent beta propriolactone (BPL) revealed that it is bound to DNA in proportion to the dose up to a definable ceiling level. It was further shown that the binding involved alkylation of the guanine component of the DNA molecule in the 7 position. At dose levels that will initiate tumors, about 1 percent of the available guanine is modified. Since the degree of binding demonstrable diminishes with time, whereas the effect is permanent, the irreversibility of biological effect must lie in the perpetuation of errors in genetic coding passed on to daughter cells, rather than in alkylated guanine itself.

Further experiments showed that DNA alkylated *in vitro* by BPL was only 50 percent effective in priming RNA synthesis systems, and that DNA isolated from mice treated *in vitro* with BPL was 85 percent effective. Furthermore, that DNA which was still effective resulted in a changed ratio of bases in the RNA synthesized. It is postulated that it is these changes in the coding of the DNA that lead to altered enzyme patterns, and thus to somatic mutations, and susceptibility to the action of promoters. If it were not for the high degree of inactivation of DNA, as compared with a relatively small degree of mutational change, carcinogenesis in response to environmental promoters would surely be much more common.

It also must be recognized that spontaneous errors of DNA replication, similar to those brought about by carcinogenic initiators, are entirely possible, and could account for a low background of nonenvironmental cancers.

---

[1] Synopsis of presentation, the substance of which appears elsewhere.

The work reported at the Symposium is discussed in the following publications:

BOUTWELL, R. K. (1964). Some biological aspects of skin carcinogenesis. *Prog. Exptl. Tumor Res.* **4**, 207–250.
COLBURN, N. H., AND BOUTWELL, R. K. (1966). The binding of β-propiolactone to mouse skin DNA *in vivo;* its correlation with tumor initiating activity. *Cancer Res.* **26**, 1701–1706.
COLBURN, N. H., AND BOUTWELL, R. K. (1968). The *in vivo* binding of β-propiolactone to mouse skin DNA, RNA, and protein. *Cancer Res.* **28**, 642–652.
COLBURN, N. H., AND BOUTWELL, R. K. (1968). The binding of β-propiolactone and some related alkylating agents to DNA, RNA, and protein of mouse skin; relation between tumor initiating power of alkylating agents and their binding to DNA. *Cancer Res.* **28**, 653–660.
BOUTWELL, R. K., COLBURN, N. H., AND MACKERMAN, C. C. (1969). In vivo reactions of β-propiolactone. *Annals N. Y. Acad. Sci.* **163**, 751–764.

## Commentary

### P. N. MAGEE

*Toxicology Research Unit, Medical Research Council Laboratories, Carshalton, Surrey, England*

Caution should be observed in extrapolating from the experiments on mouse skin discussed by Boutwell to other organs and species. The two-step process described need not always be necessary; there are agents, such as dimethylnitrosamine and N-nitrosomethylurea, that can produce tumors after a single injection without the introduction of any promoter.

Modification of intraceullar components other than DNA should not be forgotten as possibly contributing to carcinogenesis. Alkylation, moreover, may not be the only reaction involved. There are many similarities between the DNA changes in somatic cells discussed by Boutwell and those produced by some mutagenic agents in germ cells.

# Effects of Environmental Agents at the Level of Enzyme-Forming Systems

EMMANUEL FARBER[1]

*Department of Pathology, University of Pittsburgh, Pittsburgh, Pennsylvania*

The ultimate control of the environmental hazards in man's environment and the ultimate attainment and preservation of the optimum state of health would logically depend only on the identification of the hazards and their removal. Although this goal may be obtainable in many instances, the realities of modern societies suggest that economic and political considerations not infrequently thwart this tactical approach to the problem. Therefore, alternate approaches should be encouraged if the attainment of an optimal environment for man's life on earth is a desirable objective. An alternate approach, based on sound theoretical principles, is to understand, at all levels of organization, the mechanisms of the pathologic alterations induced by known environmental hazards. Such knowledge may enable man to protect himself against the harmful effects of these hazards despite his continuous exposure to them. Although knowledge about the reaction patterns of the living organism at many difficult levels of organization may suggest various ways of protection, an understanding of the reaction patterns at the cellular and molecular levels offers new possibilities at a fundamental level.

Of the many phases of cell structure and function in which progress has been made in recent years, perhaps the most exciting and most fundamental concerns the regulation of gene action and the mechanism of its phenotypic expression in protein synthesis. It used to be thought that the gene was essentially nonfunctional once a cell had become differentiated, and that the regulation of the physiologic activities of a differentiated cell is localized mainly at the level of protein synthesis. Work during the last 15 years or so, mainly in bacteria and other microorganisms, has revealed the basic fallacy of this notion. In such biologic systems, the bulk, if not all, of the genetic material, the DNA, is active throughout the life of the organism and can be evoked or suppressed by variations in the environment.

Whether this potentially also exists in most cells of higher organisms is not clear. Unlike the situation in microorganisms, much of the DNA in most highly differentiated cells appears to be both unexpressed and unavailable under physiologic conditions, presumably because of firm interaction with histones and other nuclear proteins. However, even in such cells, a certain amount of the DNA is actually or potentially available and can be altered by changes in the environment.

The only method of expression so far known for DNA as genetic material is the manufacture, in its own image, of RNA molecules (transcription) which can actively participate in protein synthesis (translation). Three major classes of RNA are now recognized—ribosomal with two subclasses (23–28S and 16–18S),

[1] Present address: Fels Research Institute, Temple University School of Medicine, Philadelphia, Pennsylvania.

transfer RNA (some 50-odd species, one or more for each of the 20-odd naturally occurring amino acids), and a spectrum of messenger RNA molecules. The ribosomal and transfer RNA molecules are part of the machinery of the cell whereby a wide variety of proteins composed of polypeptide chains can be synthesized with specific amino acid sequences. The sequence of amino acids depends on the nucleotide sequence in the messenger RNA. In this model, a small amount of nuclear DNA is available, depending on the physiologic state, for synthesis of all three types of RNA via RNA polymerase. The chemical composition of the RNA depends on the nucleotide sequence of DNA, each purine or pyrimidine of which specifies a selective pyrimidine or purine. For example, an adenine in the DNA specifies a uracil in the RNA, a guanine, a cytosine, etc. These activities reside in the nucleolus and other parts of the nucleus. In some unexplained fashion, ribosomal RNA and protein, transfer RNA, and messenger RNA pass from the nucleus to the cytoplasm where they form a specialized structure for the manufacture of protein—the polysomes. The current notion is that the polysome is composed of a strand of messenger RNA attached to which are many ribosomes. During peptide chain formation, the amino acid sequence is determined by the nucleoside sequence in the messenger, and the peptide bonds are formed by interaction of all three types of RNA on this polysome. The enzyme-forming system, therefore, consists of these components organized in specific architectural arrangements.

My responsibility in this symposium is to discuss the effects of environmental hazards at the level of enzyme-forming systems. It is clearly evident from work during the last several years that many environmental agents have a profound influence on the transcription-translation apparatus at different sites and in different directions. One group of compounds, perhaps best exemplified by methylcholanthrene and phenobarbital, act on many cells to accelerate, in specific and directed ways, the operation of this apparatus and thereby produce enhanced cellular metabolic activity. Dr. Conney is a pioneer in research with these compounds, and I shall defer to him for discussion of them.

Another group of agents act, in the opposite direction, to inhibit the operation of the system at various points, thereby altering the metabolic activity and its control by the cell. I have selected, from the many such agents, three typical ones, each of which acts on the transcription-translation apparatus at a different locus and in a different way. Each one either is an important hazard for man or is typical of such agents. The three selected are the mycotoxin aflatoxin, the chemical toxin carbon tetrachloride ($CCl_4$), and the amino acid analogue ethionine.

## AFLATOXINS

The aflatoxins (21), among which aflatoxin B is one of the commonest and most potent, are synthesized by some strains of *Aspergillus flavus* and are widespread in nature. In terms of dosage, they are a group of the most potent of all known chemical carcinogens. Aflatoxin is known to be toxic to a variety of animal species and to induce liver cancer and, to a lesser degree, cancer in some other organs. It is most probable that one or more of the aflatoxins play an important role in the induction of cancer in rats fed choline-deficient diets. They also play a major role in the development of liver cancer in trout, a disease found to be important in several fish hatcheries in the United States and elsewhere, where new commercial diets have been used during the last 10 years or so. Aflatoxins have been particularly important as a contaminant of peanuts and peanut meal. They *may* play a role in the high incidence of liver cancer in some countries in Africa.

The administration of aflatoxin is followed within a few hours by liver-cell necrosis (5, 7). This effect has been found in rats and in several other species. When fed at a very low dose (1 to 2 parts per billion of diet), it induces ductular proliferation, nodular hyperplasia, and liver cancer (12). The pattern of cellular reaction to aflatoxin in the liver is similar to that seen with many hepatic carcinogens.

Aflatoxin rapidly interacts with liver DNA to change its absorption spectrum and to inhibit its function as a template for the synthesis of RNA in the nucleus, probably including ribosomal and messenger RNA (17). Thus, this environmental agent acts on the enzyme-forming systems at the initial site of synthesis of the machinery essential for protein synthesis. In this respect, it resembles very closely the potent antibiotic actinomycin D.

The role of the interaction of aflatoxin with DNA in the genesis of acute liver-cell necrosis and in liver cancer is not known. However, it is unlikely that this molecular action of aflatoxin is responsive for *all* the inhibition of protein synthesis seen within hours after its administration or for the acute liver-cell death. Actinomycin D is just as potent an inhibitor of RNA, yet does not induce liver-cell necrosis and induces less inhibition of protein synthesis than does aflatoxin. It is probable, therefore, that aflatoxin has additional metabolic effects in the liver cell which are related to cell necrosis. It will be shown later that inhibition of RNA and protein synthesis, *per se,* even of a severe degree, is insufficient to induce liver-cell necrosis.

The effects of aflatoxin on nuclear DNA and on RNA synthesis are followed by a characteristic disorganization of the nucleolar components, as seen with the electron microscope (4, 16, 18). The nucleolus is composed of a complex strand or strands of fibrillar material intermixed in some as yet unknown manner with a granular component and with DNA. This complicated cell organelle is known to play a major role in the formation of cytoplasmic ribosomes and also probably of messenger RNA. Within a short time (60 minutes or less) following the administration of aflatoxin, one observes a physical separation of the different components of the nucleolus such as to segregate the fibrillar, granular, and DNA components in separate compact masses. These nucleolar effects of aflatoxin are virtually identical with those seen after actinomycin D. Also, other toxic environmental agents, such as the senecio alkaloid lasiocarpine, appear to have effects similar to those seen with aflatoxin and atcinomycin (6, 18). The senecio alkaloids are widespread in nature and are potent hepatic carcinogens.

Thus, one group of environmental hazards, the aflatoxins and senecio alkaloids, appear to have distinctive effects on at least one locus of the transcription-translation apparatus of the liver cell and are also potent carcinogens. Whether these correlations are significant in the pathogenesis of liver cancer remains to be investigated.

## CARBON TETRACHLORIDE

Carbon tetrachloride, although known for many years to be a potent toxin, is still an environmental hazard to which man is exposed in industry and in the home. Experimentally, the primary focus for study of this compound has been the liver, even though its deleterious effects on the kidney, the pancreas, and possibly other organs is amply documented.

This hydrocarbon rapidly induces triglyceride accumulation in the liver (fatty liver) and also acute liver-cell necrosis. Despite the volume of research work devoted to the study of these effects, it has been only quite recently that new insight into its mechanism of cellular damage at the molecular level has been forthcoming. $CCl_4$ produces a rapid inhibition of cytoplasmic ribosomal protein synthesis (16) accompanied by alterations in the appearance of the endoplasmic reticulum (ER), as observed with the electron microscope (3). It has no apparent effect on the synthesis of RNA. Existing evidence suggests a direct effect of $CCl_4$ on the chemical structure and function of the ER as the primary event with dialation of the membranes, followed secondarily by a breakup of the polysomes into monomers and a considerable depression in protein synthesis. As detailed below, the inhibition of protein synthesis is probably the basis for the accumulation of triglyceride in the liver.

Recent evidence strongly suggests that $CCl_4$ requires metabolic transformation by microsomal enzymes for its hepatotoxic action. If microsomal drug-metabolizing enzyme activities are decreased by feeding a protein-free diet, $CCl_4$ is not very hepatotoxic. If now the activities of the enzymes are increased by the administration of such compounds as DDT, more of the hepatotoxicity is restored (11). If this increase in enzyme action is reduced by the administration of such agents as SKF525A, the hepatotoxicity is again diminished. Thus, the environment may influence the effects of $CCl_4$ on the enzyme-forming system in two ways—up and down—depending on the circumstances.

New information on the mechanism of action of $CCl_4$ suggests the possibility that the primary action is concerned with lipoperoxidation mediated by some microsomal enzyme (13). The following picture thus emerges from all these recent studies: $CCl_4$ acts primarily on the physicochemical makeup of the ER by virtue of an enzyme or enzymes in the ER which facilitate lipoperoxidation of some membrane fatty acids. Many functions of membranes, including ER, are intimately tied up with their lipoprotein nature. This membrane effect secondarily affects the attached polyribosomes to facilitate their breakup into monomers, resulting in inhibition of protein synthesis. This effect is reflected in the accumulation of triglyceride in the liver. The primary effect depends on the enzymatic lipoperoxidation by $CCl_4$ and can be significantly modified in either direction by the nature of the diet and the presence of drugs which can enhance or depress microsomal (ER) enzyme activity. Since the lipoperoxidation may very well be irreversible, such an effect could lead to a progressive interference with essential functions of the membranes, and thus to cell necrosis. That the inhibition of protein synthesis *per se* is not the major trigger for cell death is suggested by the results with ethionine. Thus, a tentative working hypothesis begins to emerge to account for some of the effects of an important environmental hazard, CCl₄, on cells at the level of the enzyme-forming system (translation). A model for the interplay of a few environmental variables in modulating the molecular action of this hazard is beginning to be developed, and may lead to new insight into the molecular pathology of the cellular damage by CCl₄. This, in turn, may suggest novel ways to prevent its deleterious effects and to control this environmental hazard.

## ETHIONINE

Ethionine is an analogue of methionine in which the S-methyl group is replaced by S-ethyl. Although it was originally thought to be a product of the synthetic ability of man, it now appears that many different bacteria, including our own gastrointestinal resident, *E. coli,* produce this compound under some environmental conditions. On the basis of current knowledge, there is no indication that this compound has any practical importance as an actual or potential hazard of man. Yet, the partial elucidation of its mode of action has given us much new insight into cell reaction patterns and cell integration, and has shown that it is a very useful model for probing many aspects of molecular pathology.

The administration of ethionine leads to the induction of fatty liver (within hours), acute pancreatitis (a few days), and liver cancer (a few months). All these effects are readily prevented by the administration of methionine. The fatty liver is due to the rapid accumulation of triglyceride. The liver shows no necrosis, even many days after ethionine administration.

Ethionine rapidly induces an ATP deficiency in the liver by substituting for methionine in S-activation and by trapping thereby a considerable percentage of the cellular adenine nucleotides (10, 15). Normally, the methyl group of methionine is used as a source of methyl groups for a whole array of compounds, including RNA, choline, epinephrine, and melatonin (methyl serotonin). The precursor compound for these methylations is S-adenosylmethionine (SAM), the product of the reaction of methionine with ATP (8, 14). Under physiologic circumstances, the rate of synthesis of SAM and the rates of transmethylation from SAM are so balanced as to result in a very low steady-state concentration of SAM, and therefore not much of the adenine moiety of ATP is tied up at any one time as SAM. Ethionine is a good substrate for activation with ATP to form S-adenosylethionine (SAE). However, the rates of transethylation from SAE are much lower than with SAM. As a result, a considerable amount of the cell ATP is tied up as SAE. Although the liver cell responds by making much more adenine nucleotides from glycine and other precursors, this rate is insufficient to compensate for the rate of trapping. Consequently, the liver ATP concentration drops to low levels and remains there for at least 24 hours.

The low ATP has three effects on protein synthesis, one of which is a direct effect on the translation apparatus in the cytoplasm, the polysome system. The polysome breakup to monomers (1) and protein synthesis is inhibited (19, 20).

The liver ATP is readily prevented or reversed by the administration of an ATP precursor, such as adenine or inosine. This characteristic is what makes this model so attractive; one can study many facets of ATP metabolism, protein synthesis, and other reactions related to ATP, such as RNA metabolism, in the intact animal in a system that can be readily reversed.

The mechanism of how low ATP alters the polysome system is not yet understood. Presumably, some factor or factors required for initiation of protein synthesis and formation of polysomes are ATP-dependent.

The immediate consequence of the inhibition of protein synthesis is accumulation of triglyceride in the liver (2). This is due to the absence of the protein carrier needed to transfer triglyceride from the liver to the blood. This same mechanism probably operates in fatty liver due to $CCl_4$ and puromycin (9).

The liver of the ethionine-treated rat has a very low ATP concentration and a severely inhibited protein and RNA metabolism; yet it shows no necrosis. These findings raise the very important and provocative question: "What does one have to do metabolically to a cell to kill it?" It should be stressed that the liver is quite sensitive to the induction of cell necrosis, since so many liver agents are capable of causing liver necrosis.

## CONCLUSION

I have discussed three agents, each of which affects the enzyme- and protein-forming apparatus at a different locus and with a different mechanism. Although the complete story has yet to be told about the cellular reaction pattern of any environmental hazard, the data reviewed today are sufficiently encouraging to warrant the prediction that the expanded study in depth of the molecular pathology of selected environmental agents may lead to new and exciting insight into the interplay between man and his environment, and may well lead to new and better ways to ensure the maintenance of health and the success of man in his continual struggle with the hazards in his environment.

## REFERENCES

1. BAGLIO, C., AND FARBER, E. (1965). Correspondence between ribosome aggregation patterns in rat liver homogenates and in electron micrographs following administration of ethionine. *J. Molec. Biol.* **12**, 466–467.
2. BAGLIO, C. M., AND FARBER, E. (1965). Reversal by adenine of the ethionine-induced lipid accumulation in the endoplasmic reticulum of the rat liver. *J. Cell Biol.*. **27**, 591–602.
3. BASSI, M. (1960). Electron microscopy of rat liver after carbon tetrachloride poisoning. *Exptl. Cell Res.* **20**, 313–323.
4. BERNHARD, W., FRAYSSENIT, C., LAFARGE, C., AND LEBRETON, E. C. R. (1965). Lesions nucleolaires precoces provoquées par l'aflatoxine dans les cellules hepatiques du rat. *Acad. Sci. (Paris)* **261**, 1785–1788.
5. BUTLER, W. H. (1964). Acute toxicity of aflatoxin $B_1$ in rats. *Brit. J. Cancer* **18**, 756–762.
6. BUTLER, W. H. (1966). Early hepatic parenchymal changes induced in the rat by aflatoxin $B_1$. *Am. J. Pathol.* **49**, 113–128.
7. BUTLER, W. H. (1965). Liver injury and aflatoxin. *In* "Mycotoxins in Foodstuffs," (G. N. Wogan, ed.), p. 175. M. I. T. Press.
8. CANTONI, G. L. (1960). Onium compounds and their biological significance. *In* "Comparative Biochemistry" (M. Florkin and H. S. Mason, eds.), Vol. 1, p. 181. Academic Press, New York.
9. FARBER, E. (1966). On the pathogenesis of fatty liver. *Gastroenterology* **50**, 137–141.
10. FARBER, E., SHULL, K. H., VILLA-TREVINO, S., LOMBARDI, B., AND THOMAS, M. (1964).

Biochemical pathology of acute hepatic adenosinetriphosphate deficiency. *Nature* **203**, 34–40.
11. McLean, A. E. M., and McLean, E. K. (1966). The effect of diet and 1,1,1-trichloro-2,2-bis-(*p*-chlorophenyl)ethane (DDT) on microsomal hydroxylating enzymes and on sensitivity of rats to carbon tetrachloride poisoning. *Biochem. J.* **100**, 564–571.
12. Newberne, P. M. (1965). Carcinogenicity of aflatoxin contaminated peanut meals. In "Mycotoxins in Foodstuffs," (G. N. Wogan, ed.), p. 187. M. I. T. Press.
13. Recknagel, R. O., and Ghoshal, A. K. (1966). Lipoperoxidation as a vector in carbon tetrachloride hepatotoxicity. *Lab. Invest.* **15**, 132–146.
14. Shapiro, S. K., and Schlenk, F. (1960). The biochemistry of sulfonium compounds. *Advan. Enzymol.* **22**, 237–280.
15. Shull, K. H., McConomy, J. M., Vogt, M., Castillo, A. E., and Farber, E. (1966). On the mechanism of induction of hepatic adenosine triphosphate deficiency by ethionine. *J. Biol. Chem.* **241**, 5060–5070.
16. Smuckler, E. A., Iseri, O. A., and Benditt, E. P. (1962). An intracellular defect in protein synthesis induced by carbon tetrachloride. *J. Exptl. Med.* **116**, 55–72.
17. Sporn, M. B., Dingman, C. W., Phelps, H. C., and Wogan, C. N. (1966). Aflatoxin $B_1$: Binding to DNA *in vitro* and alteration of RNA metabolism *in vivo*. *Science* **151**, 1539–1541.
18. Svoboda, D. J., and Higginson, J. (1966). Ultrastructural lesions in rat and monkey liver due to aflatoxin. *Federation Proc.* **25**, 361.
19. Villa-Trevino, S., Farber, E., Staehelin, T., Wettstein, F. O., and Noll, H. (1964). Breakdown and reassembly of rat liver ergosomes after administration of ethionine or puromycin. *J. Biol. Chem.* **239**, 3826–3833.
20. Villa-Trevino, S., Shull, K. H., and Farber, E. (1963). The role of adenosine triphosphate deficiency in ethionine-induced inhibition of protein synthesis. *J. Biol. Chem.* **238**, 1757–1763.
21. Wogan, G. N., (Ed.). (1965). Mycotoxins in Foodstuffs. M. I. T. Press.

## Commentary

### A. H. Conney[1]

*Department of Biochemical Pharmacology, The Wellcome Research Laboratories, Burroughs Wellcome and Co. (USA), Inc., Tuckahoe, New York*

Dr. Farber pointed out the inhibitory effects of aflotoxin, carbon tetrachloride, and ethionine on protein synthesis in the liver, and he indicated the possible *in vivo* significance of these effects. Treatment of animals with other foreign chemicals, such as phenobarbital, 3-methylcholanthrene, and chlordane, stimulate protein synthesis, resulting in enhanced liver growth and in greatly increased levels of enzymes in liver microsomes that metabolize drugs and steroid hormones. The increased enzyme levels are paralleled by enhanced drug and steroid metabolism *in vivo*. For instance, pretreatment of rats with phenobarbital or chlordane inhibits the actions of exogenously administered estradiol, testosterone, and pro-

[1] Present address: Department of Biochemistry, Hoffmann-La Roche Research Laboratory, Nutley, New Jersey.

gesterone. Recent studies suggest that inducers of liver microsomal enzymes enhance the hydroxylation of steroids in man. Phenobarbital, diphenylhydantoin, and phenylbutazone are examples of drugs that stimulate the hydroxylation of cortisol by enzymes in guinea pig liver microsomes and enhance the excretion of 6$\beta$-hydroxycortisol in human urine. Further research is needed to determine the physiological significance of the stimulatory effect of drugs and environmental chemicals on the microsomal metabolism of steroids and other normal body constituents in man. The answer to this question is particularly important because man is exposed to increasingly large numbers of drugs, pesticides, herbicides, food additives, and environmental polycyclic hydrocarbons that stimulate drug metabolism and the metabolism of normal body constituents.

# Effects of Environmental Agents at the Enzyme Level—Air Pollutants[1]

J. Brian Mudd

*Department of Biochemistry and Statewide Air Pollution Research Center, University of California, Riverside, California 92507*

Of the various pollutants present in Los Angeles air, peroxyacetyl nitrate (PAN) is the one whose action at the enzyme level has been most studied. It is an end product of a series of reactions occurring when nitrogen oxides and oxygen are irradiated in the presence of hydrocarbons. The three possible toxic modes are: oxidation due to the peroxidic character; acetylation due to the anhydride character; and liberation of nitrite on hydrolysis. Its half life in aqueous solution at pH 7 is about 5 minutes. It is most likely to have its effects, therefore, on susceptible chemical targets close to the portal of entry in the lungs. Enzyme systems as potential targets are considered here.

Nicotinamide nucleotides are readily oxidized by PAN, so that one would expect energy-producing reactions, normally mediated by NAD in the cytochrome system, to be impaired. The physiological significance, however, has not been assessed.

Glucose-6-phosphate dehydrogenase is inactivated by PAN. It is protected by NADP with which it is usually associated, but not by glucose-6-phosphate. Isocitrate dehydrogenase is protected by both NADP and isocitrate. Malate dehydrogenase is protected by neither NAD nor malate. These differences are similar to inhibition by sulfhydryl reagents and indicate that PAN inactivates enzymes by reaction with the sulfhydryl group.

Glutathione is converted by PAN to oxidized glutathione and to S-acetyl glutathione. The sulfhydryl groups of proteins, however, are relatively inaccessible, and there is steric difficulty in forming disulfide bonds. Experiments with hemoglobin show that the effect is limited with the tetrameric molecule, but that when the molecule is broken down to the constituent monomers the reaction of PAN with sulfhydryl groups increases.

Experimental work warrants the conclusion that the reaction of PAN with sulfhydryl groups produces disulfides, S-acetyl compounds, and sulfoxides. These reactions inactivate enzymes which require free sulfhydryl groups for activity. There is no inactivation of enzymes which have no sulfhydryl groups, such as ribonuclease. Reaction of PAN with glutathione may be a detoxifying mechanism under physiological conditions. Conversion of protein sulfhydryl groups to disulfides and S-acetyl compounds may be reversible to disulfide reductases and thioesterases. In these cases one would expect the inhibitory effect to be only temporary. These suggestions based on in vitro experiments need testing under

[1] Synopsis of presentation, the substance of which appears elsewhere.

physiological conditions.

The work reported at the Symposium is discussed in the following publications:

MUDD, J. B., AND DUGGER, W. M. (1963). The oxidation of reduced pyridine nucleotides by peroxyacetyl nitrates. *Arch. Biochem. Biophys.* **102**, 52–58.
MUDD, J. B. (1963). Enzyme inactivation by peroxyacetyl nitrate. *Arch. Biochem. Biophys.* **102**, 59–65.
MUDD, J. B. (1966). Reaction of peroxyacetyl nitrate with glutathione. *J. Biol. Chem.* **241**, 4077–4080.
MUDD, J. B., LEAVITT, R., AND KERSEY, W. H. (1966). Reaction of peroxyacetyl nitrate with sulfhydryl groups of proteins. *J. Biol. Chem.* **241**, 4081–4085.
MUDD, J. B., AND MCMANUS, T. T. (1969). Products of the reaction of peroxyacetyl nitrate with sulfhydryl compounds. *Arch. Biochem. Biophys.* **132**, 237–241.

## Commentary

JAMES R. GILLETTE

*Laboratory of Chemical Pharmacology, National Heart Institute, Bethesda, Maryland*

Some sulfhydryl inhibitors have paradoxical effects on the activities of the affected enzyme. Mercury compounds, for example, have differential effects upon the catalytic ability of glutamic dehydrogenase with respect to alanine and glutamic acid. Similarly, the activity of TPNH cytochrome C reductase is differentially affected with respect to cytochrome C on the one hand and to neotetrazolium and neoprontosil on the other. On the other side of the coin, a number of compounds other than sulfhydryl inhibitors can inhibit microsomal enzymes through a variety of mechanisms. Some may act as noncompetitive inhibitors at a low concentration but as competitive inhibitors at the active site at higher concentrations.

# Growth and Trophic Factors in Carcinogenesis

P. N. MAGEE

*Toxicology Research Unit, Medical Research Council Laboratories, Woodmansterne Road, Carshalton, Surrey, England*

"Autonomy of the cancer cell is a relative matter and there is an effect of the host on the tumor" (1).

The topic, "Growth and Trophic Factors in Carcinogenesis," is extremely wide and it is difficult to decide where to begin. In this discussion, an attempt will be made to examine the validity of the above statement, by R. W. Begg, and to assess the importance of growth and trophic factors in relation to cancer induced by hazards present in man's environment.

It has been known since the 18th century, from the work on scrotal cancer in chimney sweeps by Percival Pott, that cancer can be and is induced by environmental agents. There is no reason to think that this differs in any way from cancer of unknown etiology. From a practical point of view, it is important to determine to what extent, if any, the fate of such tumors is influenced by factors in the environment or in the host, rather than in the neoplastic cells themselves.

### SPONTANEOUS REGRESSION OF HUMAN CANCER (12, 13, 29)

There seems to be little doubt that apparently spontaneous regression of cancer in man can occur. There is no doubt whatever that it is a very rare event, and the closer the scrutiny, the smaller the number of acceptable cases becomes.

Everson (12) has evaluated more than 1000 cases of alleged spontaneous regression of human cancer and concluded that only 130 had been adequately studied. Regression was most commonly noted in neuroblastoma, hypernephroma, choroiocarcinoma, and malignant melanoma, none of which is common. Various factors possibly involved in regression were discussed. Endocrine influences may have been responsible in some cases of regression of breast cancer, and the palliative effect of bilateral oophorectomy, adrenalectomy, or hypophysectomy is well established in the treatment of metastatic breast carcinoma. Other suggested possible factors that may be involved included unusual sensitivity to usually inadequate therapy, fever and/or acute infection, and allergic or immune reactions, the latter being described as "a rich field for speculation." Further possibilities were interference with nutrition of the tumor, removal of a carcinogenic agent, complete surgical removal of the tumor, and incorrect histologic diagnosis of malignancy.

Even if critical appraisal shows that permanent regression of cancer does not occur, it is generally agreed that cancer cells often lie latent in tissue for long periods under natural and experimental conditions. It is reasonable to suppose that resistance of the host may be responsible for this. Evidence for host resistance to cancer in man comes from several observations (24). Successful tumor autografts are rare, although the original tumors progress. Tumor cells have been frequently demonstrated in the circulating blood, and some must settle in the spleen, but metastatic deposits are very rare in this organ. Spontaneous regression of tumors probably occurs, as discussed above, and also metastases appear to regress after removal of the primary tumor and others may remain latent for long periods.

The following presentation will be concerned with some of the factors which may be involved in controlling the incidence and rate of growth of tumors, mainly in experimental animals. An attempt will then be made to assess the significance of these factors in relation to health hazards in man's environment. Finally, some of the conclusions will be considered in relation to carcinogenesis by certain nitrosamines and related compounds which are very potent carcinogens and may constitute an environmental hazard. Most of the presentation will be taken from the published work of others and the references, with few exceptions, will be to review articles.

The factors to be discussed which possibly affect the incidence and growth of tumors will be dose-response relationships for carcinogens, initiation and promotion in carcinogenesis, progression of tumors, dependence and autonomy of tumors, and hormonal, immunologic, and nutritional factors. As is well known, the genetic background of the host is of extreme importance in many forms of carcinogenesis, but this will not be discussed.

## DOSE-RESPONSE RELATIONSHIPS IN CARCINOGENESIS (6, 8, 10)

These are extremely complex and will be discussed only briefly. There is evidence that the number of tumors produced and the latent period of their induction is dependent on the dose and duration of treatment with chemical carcinogens. The question of threshold levels for the induction of cancer is extremely important from the standpoint of environmental carcinogenesis. Dose-response curves with some chemical carcinogens appear to be linear according to Boyland (6) who, with others, maintains that no threshold or safe level therefore exists. Similar conclusions have been drawn from work in radiobiology; it has been stated by the United Nations Scientific Committee on the Effects of Atomic Radiation (1962) that "biological effects follow irradiation, however small its amount."

## TWO-STAGE THEORY OF CARCINOGENESIS; TUMOR PROGRESSION; DEPENDENT AND AUTONOMOUS TUMORS

The origin of these three concepts can be traced back to the work of Rous and Kidd (40), who induced papillomas in the skin of rabbits' ears by repeated applications of tar. The tumors induced were nearly always benign and some time elapsed before cancer appeared, frequently arising from benign growths. These benign tumors were usually numerous and often very large, but their pathologic classification was not immediately clear and they were lumped together under the general description of warts. These lesions are more suitable for study in rabbits than in mice because they retain their initial character longer and may become malignant only after months or years and in only a small proportion of cases. The cells of these lesions have an abnormal capacity for proliferation, which is an attribute of neoplasia, but, as emphasized by Rous and Kidd, this capacity is exerted only under favoring conditions. Unless the applications of tar are kept up or have already induced chronic local changes, all the warts dwindle or disappear, except those few which become cancers. These warts of rabbits are tumors by all standard criteria but two: They have no capacity for independent growth in the absence of the carcinogenic stimulus, and the changes in their cells may be reversible since they become smaller and vanish. Reapplication of tar after an interval led to an increased response, and it was shown to stimulate proliferation of the warts, as well as to induce them. Warts which disappeared left no discernible trace, but those which reappeared at the site of previous lesions in response to renewed tarring were not new lesions but recurrences, because their individual characteristics were always similar to those previously present at the same sites. A further, highly significant observation was that recurrence of warts did not necessarily require further application of the carcinogen, since certain noncarcinogenic stimuli were shown to provoke recurrences. Among these stimuli were turpentine applied to areas of ear skin where warts had disappeared, and wound healing at the site of punch-holes in the ears where biopsy specimens had been taken. Application of tar to the site where lesions had been promptly gave rise to warts as late as 6 months after the previous application had been discontinued. There is thus no doubt that, although the epithelium of regressing papillomas has a normal microscopic appearance, it must retain the essential potentialities of tar wart epithelium.

The fate of the majority of these papillomas is interesting, because even some growths which proliferate vigorously for many months and retain distinctive characters finally prove unable to maintain themselves. These tar warts are thus conditional in nature and are wholly dependent on aid for continuing survival.

This paper of Rous and Kidd, which is one of the classics of experimental pathology, has been considered in some detail because it contains so many ideas that have been subsequently developed by others and have now become a part of the fabric of our knowledge of cancer. It clearly foreshadows the two-stage theory of carcinogenesis, the idea of tumor progression, and the whole concept of the dependent or conditioned tumor. Each of these topics will be considered briefly in turn.

Two-Stage Theory of Carcinogenesis and Co-Carcinogenesis (2, 3, 41, 42)

These ideas have already been discussed by Dr. Boutwell and will be treated only in summary fashion here. The term "co-carcinogenesis" was introduced by Murray Shear in 1938 to describe the enhancement of the carcinogenic action of 3,4-benzpyrene by a basic fraction of creosote. It has subsequently been used to cover a wide range of phenomena in which tumors are produced by one or more agents, acting together or serially.

The foundation of the two-stage theory was laid by Friedewald and Rous (18), who introduced the terminology which has now been universally accepted. Rous and his colleagues, including Beard and McKenzie, worked on the skin of rabbits, and other investigators—notably, Mottram, Berenblum, Shubik, Tannenbaum, and Rusch—carried out similar experiments on mouse skin. All agreed that the process of carcinogenesis was divisible into stages. It was suggested that there is an initial stage, involving conversion of some normal cells into "latent tumor cells" or sensitized cells, and that these, after a latent period, began to multiply and form tumors during a second stage. These two processes have been given a variety of names, but the generally accepted terminology is that of Friedewald and Rous (18), which is "initiation" and "promotion" for the two stages and "latent tumor cells" for the altered cells formed during the first stage.

The results of many experiments of this kind were expressed in a rather precise way by Berenblum and Shubik (4), who described a "model experiment" which has, in fact, been a model for much subsequent work. A single dose of a carcinogen (the initiator) was applied to the skin of mice, followed, after varying intervals, by repeated applications of croton oil (the promoter). The average latent period from the beginning of the promoting stimulus to appearance of the tumors did not change with varying the dose of carcinogen or the time between initiation and promotion. The proportion of tumor-bearing mice rose to a "set-level," less than 100%, which was dependent on the initiating dose of the carcinogen but independent of the interval between initiation and promotion. The initiating and promoting stimuli used induced only an occasional tumor when applied separately or in the reverse order. Berenblum and Shubik concluded that an initiating stimulus converted a relatively small number of normal cells into latent tumor cells, and that this change is irreversible. Only these cells give rise to tumors under the influence of the promoting stimulus.

## Progression of Tumors (14–17, 25)

The concept of tumor progression has been put forward by Leslie Foulds in a series of papers appearing over the last three decades. His basic conclusion is that neoplasia is discontinuous in space and time—that it is a dynamic process advancing through stages that are qualitatively different. Progression is defined as an irreversible qualitative change in one or more of the characters of neoplastic cells.

Foulds has formulated certain principles of tumor progression based on analysis of his own studies and those of many others reported in the literature. The systems that he mainly considered were chemically induced tumors of the skin, hormone-induced tumors, the spontaneous mammary and uterine tumors of the rabbit studied by H. S. N. Greene, and two types of tumor which he had studied in detail himself. These were tumors of the urinary bladder induced by acetylaminofluorene and spontaneous mammary tumors, both in mice.

*Independent progression of tumors.* This means that progression occurs independently in different tumors in the same animal. Foulds based this conclusion on the observation that only one of the multiple mammary tumors in a mouse undergoes progression at a time. The same rule applies to chemically induced tumors of mouse skin and bladder. It is also consistent with the well-known clinical observation in man that only a few of the numerous papillomas in familial intestinal polyposis will become carcinomas and only one carcinoma usually arises in the skin lesions of shale oil workers.

*Independent progression of characters.* Progression occurs independently in different characters in the same tumor. This rule applies to a number of different characters exhibited by a tumor—such as growth rate, histologic type, invasiveness, and power of metastasis—which Foulds has called "unit characters" of the tumor. He concludes that these "unit characters" are independently variable within wide limits, but that they can be assorted and combined in a variety of ways. Among the various unit characters of a tumor, its responsiveness to hormones is of great importance in relation to control of its growth and will be discussed in more detail below.

*Progression is independent of growth.* This rule is derived from observations on mammary tumors in mice which have regressed after parturition but recur without the stimulus of pregnancy, indicating that they have progressed from a responsive to an unresponsive condition without the necessity of growth. It also appears to apply to those tumors in man which progress to an uncontrollable state while their growth is checked by hormones or chemotherapy. It is well known that tumors can be kept in remission for long periods by chemotherapy, only to enter a terminal phase of rapid growth and dissemination. Two corollaries of this rule are that, when first observed, tumors may be at any stage in progression, and that progression is independent of the size and clinical duration of the tumor.

*Progression is continuous or discontinuous.* This simply means that progression may occur by gradual change or by abrupt steps, and is also derived from observations of a variety of neoplastic conditions.

*Progression follows one of alternative paths of development.* This expresses the observation that tumors may follow different paths leading to the same or different end points. There are many examples of this which underline the essential unpredictability of the course of most tumors.

*Progression does not always reach an end-point within the lifetime of the host.* This rule is illustrated by the behavior of many tumors during serial transplantation. For example, hormone-induced tumors in rodents often remain dependent on hormones in their primary hosts, and chemically induced skin tumors rarely advance far in rabbits; the same applies after brief exposure to the carcinogen in mice. An obvious example from human pathology is the well-known behavior of the so-called latent prostatic carcinoma.

Clearly, it is not possible to discuss the work of Foulds at greater length, and the reader is referred to his various review articles for further information.

However, he has drawn certain conclusions which are extremely important in any consideration of growth and trophic factors in carcinogenesis. Foulds believes that, for many forms of neoplasia, initiation by the carcinogen seems to establish a region of "diffuse incipient neoplasia" from which tumors develop later by focal progression. This view is quite consistent with that of R. A. Willis, based largely on human autopsy material, that cancer develops in regions or "fields" of tissue which have become diffusely "predisposed to" or "prepared for" neoplastic change. Progression is not dependent on continued cell proliferation and seems to occur without any identifiable stimulus. Finally, and most significantly, Foulds concludes that the quality and duration of the inducing stimulus seem to determine the potentialities for subsequent progression, and its occurrence within the region of potential neoplastic change is apparently random in time and in site. If this is correct, the implication seems to be that the ultimate fate of most tumors is determined at a very early stage and that, after this, little can be done to influence the course of events. The importance of this conclusion, if true, for environmental carcinogenesis is obvious.

However, it is well established that the course of some tumors can be influenced, and this leads to the next topic for discussion, the concept of the conditioned or dependent tumor.

### Autonomous and Conditioned or Dependent Tumors (19, 20, 27, 40)

This concept has already been mentioned in reference to the work of Rous and Kidd on skin tumors and Fould's work on responsive and unresponsive mammary cancer in mice. It has been extensively studied by Greene using the technique of tumor transplantation into the anterior chamber of the eye in rabbits.

The term "conditioned tumor" was introduced by Rous and Kidd (40) to describe the skin papillomas induced in rabbits by tar which regressed when the application of the carcinogen was discontinued. Another example, from the work of Foulds, is the dependence of some mammary tumors in mice on pregnancy. These tumors grow only during pregnancy, reach a peak at parturition, and then regress partially or completely. In this case, the conditional character of the tumors is clearly hormone-dependent and it is in the field of hormonal carcinogenesis that conditional neoplasia is most often encountered. Hormonal factors in carcinogenesis will now be considered.

### HORMONAL FACTORS IN CARCINOGENESIS (5, 9, 17, 19–24, 34–36)

As is well known, hormones play an important role in the etiology of certain forms of cancer and they can also exert marked effects on fully developed tumors. Beatson, in 1896, reported the value of ovariectomy in women for the alleviation of advanced breast cancer. This procedure was used intermittently for about 10 years before going out of favor, and it was not until 1941 that Charles Huggins and his colleagues reported the favorable results of orchidectomy on advanced cancer of the prostrate. On the experimental side, Lathrop and Loeb, in 1916, found that ovariectomy in early life reduced the incidence of spontaneous mammary tumors in mice and Lacassagne showed, in 1932, that mammary tumors could be induced by prolonged administration of ovarian hormones in male mice of a strain with a moderately high incidence in females but negligibly low in males. Subsequently, of course, a variety of endocrine derangements have been shown to induce neoplasia, not only in breast, but also in other organs, including the gonads, adrenals, pituitary, and thyroid. The derangements of endocrine homeostasis have included administration of estrogens, gonadectomy, irradiation, and the grafting of endocrine organs into abnormal anatomic situations. Genetic factors often play an important role in hormonal carcinogenesis.

According to Mühlbock, all hormones which have a carcinogenic effect have the common property of stimulating growth in particular target organs, which, with few exceptions, include the primary endocrine organs and the secondary sex organs. Thus, tumors of the pituitary are induced by estrogens and progesterone; of the thyroid, by thyrothropin; of the adrenals, by adrenocorticotropin; and of the ovary and testis, by gonadotropins. In the secondary sex organs, tumors of the uterus are induced by estrogens, and of the mammary gland, by mammotropic hormone. The exceptions include the induction with estrogens of leukemia in mice by Kaplan, and of kidney tumors in the male golden hamster by Kirkman and Horning.

Other hormones, such as insulin and thyroxin, which have a general action on metabolism without specific target organs, are now known to be carcinogenic. The growth hormone of the pituitary also has no special target organ and it is uncertain whether it is, by itself, carcinogenic. There is evidence, however, that it can probably stimulate the growth of chemically induced tumors of the liver which will be discussed later.

The carcinogenic effect of hormones is always dependent on excessive action on the target organ, and hormone deficiency does not give rise to neoplastic growth in the target organ, which becomes atrophic. Mühlbock and his colleagues have shown that the carcinogenic effect is related to the amount of hormone acting on the target organ; that is, there is a dose-response relationship. Furthermore, tumors will develop only under continuous growth stimulation by the hormone of the target organ, and intermittent administration of hormone is not effective. These quantitative relationships were demonstrated for the induction by estrone of pituitary tumors in mice and mammary tumors in gonadectomized mice. Cancer cannot be induced by a single large dose of hormone, and the tumor-inciting action of hormones is not irreversible before tumor formation.

It is quite clear, as Mühlbock has pointed out, that the induction of cancer by hormones differs in many ways from induction by chemical carcinogens, and these differences have been emphasized by Jacob Furth, whose contributions to this field have been enormous. Furth (19, 23) contrasts the mechanisms of chemical and hormonal carcinogenesis and concludes that the processes are fundamentally different. He accepts the point that chemical carcinogenesis probably involves an initial alteration at the cellular level, probably involving some change in the hereditary material, but maintains that such a change cannot explain the origin and character of all cancers. Many tumors appear to arise from disturbance of the homeostatic equilibria which regulate the numbers of each cell type; the result of the disturbance is progressively growing tumors. According to Furth, the basic change is not in the cells to be regulated, but in the regulatory apparatus, and this change can be reversed. In other words, cancer can originate from a primary change in the host, as well as from a primary change in the cell which is to become neoplastic. This idea is derived from studies on dependent or conditioned and autonomous neoplasms, which will now be considered in more detail.

The concept of conditional tumors has already been discussed in relation to the work of Rous and Kidd on chemical carcinogenesis of the skin, but now requires redefinition in relation to hormonal carcinogenesis. The following definitions are due to Clifton (9).

(1) *Dependent* neoplasms will grow only in specifically modified or conditioned hosts. They will not grow when grafted in normal histocompatible animals but, when grafted in conditioned hosts, they invade locally, can metastasize, and are eventually fatal.

(2) *Autonomous* neoplasms will grow in unconditioned animals. They often develop from dependent neoplasms of endocrine glands, may be highly secretory and often are responsive in that they are stimulated by specific physiologic factors.

Furth emphasizes that the capacity for unrestrained proliferation is not a distinguishing feature of cancer cells, because normal cells can proliferate without restraint in tissue culture. Obviously, there must be restraining forces in intact organisms which keep the cells within normal bounds and there must also be specific stimulating influences which can regulate the numbers of cells in the different organs. All cells are under the control of physiologic homeostatic mechanisms to a greater or lesser extent. If these homeostatic mechanisms are deranged toward stimulating influences, tumors may result. This has been well illustrated by experimental derangements of several homeostatic systems, including the pituitary-thyroid, the pituitary-gonadal, and the pituitary-mammary gland systems.

## Pituitary-Thyroid System

This is one of the best situations for demonstration of tumorigenesis by disturbance of homeostasis, since tumors of either organ can be induced by the appropriate experimental procedure and they can be fully controlled, in their earlier stages, by restoration of the normal homeostatic equilibrium.

Tumors of the thyroid can be induced by sustained interference with thyroid hormone production by the administration of goitrogens, such as thiouracil. This treatment causes reduction in thyroid hormone production without damage to the proliferative capacity of the thyroid cells. This causes a fall in the level of the circulating thyroid hormone, which, in turn, causes increased secretion of the thyrotropic hormone of the pituitary. The continued action of the thyrotropic hormone on the thyroid leads to hyperplasia, followed by adenoma and finally carcinoma of the thyroid. Indistinguishable thyroid tumors can also be induced in normal mice subjected to excessive quantities of thyroid-stimulating hormone by grafting of pituitary tumors which secrete this hormone. The tumors induced by goitrogens have been extensively studied by Bielschowsky and others. They may have the histologic features associated with malignancy and they can invade lymphatics and metastasize to the lungs. On transplantation, however, they will grow only in mice whose synthesis of thyroid hormone has been blocked—for example, by thiouracil. They will not grow in mice with a normally functioning thyroid. Thus, although these tumors have many of the characteristics of malignancy, they depend on the continued presence of thyrotropic hormone for their survival and growth. It is important to note, however, that, in the course of repeated transplantations in thiouracil-treated mice, the character of these tumors changes and they become capable of growth in mice with normal thyroid function. In the terminology of Foulds, they have progressed from hormone dependence to autonomy.

By another experimental procedure, tumors of the thyrotropic hormone-secreting cells of the pituitary can be produced. In this procedure, the thyroid of the mouse is destroyed by localized radiation injury induced by large doses of radioactive iodine ($^{131}I$), of the order of 200 $\mu$Ci, advantage being taken of the selective uptake of iodine by the thyroid. In contrast to the action of thiouracil, the radiation injury produces atrophy of the thyroid, which therefore becomes incapable of hyperplasia. The resulting deficiency of thyroid hormone causes sustained stimulation or loss of restraint of the pituitary, which leads to the formation of multifocal pituitary adenomas. The induction of these tumors by so-called radiothyroidectomy can be prevented by thyroid grafts or administration of thyroid hormone. These pituitary tumors can be grafted on to mice that have been subjected to radiothyroidectomy to depress thyroid function and will grow, metastasize, and be as lethal as other cancers. They cannot be grafted on to hosts with normal thyroid function. As with the thyroid tumors, however, these hormone-dependent or conditioned neoplasms frequently give rise, on serial transplantation, to autonomous cancers which will grow in normal hosts and cannot be restrained by thyroid hormone.

## Pituitary-Gonadal System

Ovarian tumors can be induced by grafting ovaries into the spleen or other sites drained by the portal vein in castrated animals, including rats, mice, rabbits, and guinea pigs. This is because the liver destroys the ovarian hormones, leading to a disturbance of homeostasis and hypersecretion of pituitary hormone. Testicular neoplasms, including Leydig's cell tumors and teratomas, have been induced in castrated rats by intrasplenic grafts of testis. The ovarian tumors are predominantly granulosa cell tumors with occasional luteomas and they can give rise to metastases in the liver. This does not necessarily mean that the tumors are autonomous, and the results of transplantation experiments into normal hosts have been inconclusive. Injection of gonadal hormones into mice bearing intrasplenic ovarian grafts did, however, prevent the development of ovarian tumors.

Chromophobe adenomas of the pituitary can be induced by estrogens, and the incidence is reduced by androgen administration. Such tumors will grow when transplanted into mice of the same strain if the hosts are given estrogen, but do not grow in normal mice. They can progress to grow in the absence of estrogen.

## PITUITARY-MAMMARY GLAND SYSTEM

Studies on this system have thrown valuable light on the interrelationships of hormonal factors in viral, chemical, and radiation-induced carcinogenesis. They also illustrate the interplay of carcinogens, hormones, and genetic factors in the induction of tumors. Furth and his colleagues have carried out quantitative studies on the induction of mammary tumors in mice by radiation, viral, and chemical carcinogens, and have found that in each case the mammotropic hormone has marked effects on the incidence and growth of the tumors. In these experiments, the hormonal effect was obtained by the implantation of functional mammotrope tumors of the pituitary induced by irradiation. These grafts cause a marked stimulation of the mammary gland. With low doses of radiation (50 R) or mammotropic hormone alone, no tumors were obtained and only a small proportion of rats given a higher dose of radiation (150–400 R) developed tumors. When rats given the lower dose of radiation also received the mammotropic stimulus, the incidence of tumors rose to over 50%, and with the larger dose of irradiation the incidence was higher still. Since these experiments were done in the rat, there was no need to implicate a virus, but in experiments on tumor induction in mice by the mammary agent of Bittner, a similar clear-cut effect of the mammotropic hormone was observed. Perhaps the most interesting of these experiments were those on the effect of the hormone on chemical induction of mammary cancer in the rat. As Charles Huggins has shown, the most consistently successful method of inducing mammary neoplasia in experimental animals is to introduce a polycyclic hydrocarbon carcinogen into the stomach of the rat. Furth and his colleagues induced palpable mammary tumors in three groups of female rats by repeated feeding with 3-methylcholanthrene. The one group was left as a control, the second was ovariectomized, and the third hypophysectomized. In two-thirds of the ovariectomized and thyroidectomized rats, the tumors decreased in size, and in some they regressed completely. When mammotropic hormone was administered to these rats by implantation of the mammotrope pituitary tumor, growth resumed in the atrophic tumors and many new tumors also appeared. When the methylcholanthrene was given in lower doses, no tumors appeared, but mammotropic stimulation before or after the chemical carcinogen caused the appearance of mammary tumors in most of the animals. It is interesting that the hormonal treatment could be delayed for as long as 8 months after administration of the chemical carcinogen and still cause mammary tumors. The similarity of this situation to that in skin carcinogenesis is obvious.

## EFFECTS OF HORMONES ON TUMORS IN ORGANS WHICH ARE NOT SPECIFIC TARGETS OF HORMONAL ACTION

The hormones to be considered are corticosteroids and the pituitary growth hormone or somatotropin.

*Corticosteroids.* Cortisone often has a strongly inhibiting influence on the growth of tumors, but this effect is not restricted to hormone-dependent tumors and may be related to the unfavorable general condition induced in the tumor-bearing animals. This may be compared with nutritional effects on tumor growth. A direct effect of cortisone on the tumor cell cannot be excluded.

*Somatotropic hormone.* As mentioned earlier, the growth hormone of the pituitary has no specific target organ, but it has been shown to influence experimental liver carcinogenesis (44, 48). There is evidence that hypophysectomy and thyroidectomy abolish the carcinogenic action of azo dyes and acetylaminofluorene (or fluorenylacetamide) on rat liver, but have no action on the induction of carcinoma of the ear-duct by the latter carcinogen. Administration of pituitary hormones partially restored the liver carcinogenicity. Weisburger and his colleagues attempted to demonstrate a direct involvement of the pituitary in liver carcinogenesis by the use of one of the transplantable functional pituitary tumors developed by Furth. This tumor is reported to be rich in growth hormone, ACTH, and prolactin, but generally to secrete little gonadotropin. Implantation of this tumor was found to reduce the time of induction of liver tumors by N-hydroxyfluorenylacetamide in the rat, the first histologically recognizable tumors appearing as early as 13 weeks after the start of carcinogen feeding. Implantation of the pituitary tumors also reinstated the hepatocarcinogenic effects of fluorenylacetamide in hypophysectomized and thyroidectomized rats, suggesting the possibility that growth hormone is the significant agent. Weisberger claims that these findings support the concept that cancer formation in liver requires the action not only of a chemical carcinogen, but also of pituitary hormones.

## General Comments

There is no doubt that hormones may exert an important influence on the induction and subsequent fate of tumors. Induction of tumors by hormones differs in several ways from induction by chemical carcinogens, suggesting that the hormonal effect may be more that of a promoter than that of an initiator. Hormones appear to induce neoplasia by proliferation of fully differentiated cells, and the tumors so induced usually pass through a phase of dependence on continued hormone stimulation which varies in duration. With few exceptions, these tumors occur in the specific target organs of the hormones—namely, the endocrine and secondary sex organs. All hormone-dependent tumors show the phenomenon of progression, as defined by Foulds, and sooner or later become autonomous. There may be a change in tumor morphology, but the change to autonomy may occur without visible structural changes. It is usually characterized by an increase in growth rate. The change from dependence to autonomy is unpredictable, and the factors which influence it are unknown. In spite of this, the control of carcinoma of the prostate by endocrine therapy introduced by Charles Huggins is still among the most effective methods of treatment available for human cancer.

## IMMUNOLOGIC FACTORS IN CARCINOGENESIS (7, 29, 33, 37, 51)

The probable regression of some human cancers and the undoubted evidence of tumor resistance in man have been discussed above. The nature of tumor resistance has led to much speculation in the past, and the possibility of an immunologic mechanism was raised many years ago. As Furth (24) has pointed out, there have been three eras of cancer immunology.

The first began in 1906, with the discovery by Paul Ehrlich that tumor transplants often failed to take in mice which had previously received a transplant of the same or another tumor. This observation was subsequently confirmed by many other workers, but, unfortunately, due largely to the lack of genetically uniform strains of laboratory animals, it was not realized that rejection was not due to any intrinsic property of the tumor, but to what is now recognized as the homograft reaction. This period came to an end in 1929 with the publication of a highly critical review by Woglom.

The second era was made possible by the development of in-bred strains of animals initiated by Little at Bar Harbor. With the aid of these animals, Snell, Gorer, Medawar, Mitchison, and others were able to carry out the precise immunogenetic analyses of transplantation of neoplastic and normal tissues which have led to our knowledge of the homograft reaction and immunologic tolerance. It is now known that transplants, whether of normal or neoplastic tissues, from one animal to another of the same species provoke a state of immunity in the recipient. As a result of this, the transplants are sooner or later destroyed. The antigens concerned are genetically determined by the histocompatibility genes. Immunization does not occur on transplantation of normal tissues between in-bred strains because the recipient possesses all the antigens present in the donor. Even when the donor does possess antigens which are not present in the recipient, the graft may not be destroyed if the antigens are weak, if the immunologic response of the recipient is weakened, or, as often happens with tumor transplants, if the transplant can survive and continue to grow in spite of a powerful immunologic reaction. It is sometimes possible to demonstrate antibody in the serum of homograft recipients. This antibody may play a role in destruction of the graft, but usually cellular attack is of major importance. On the other hand, the presence of antibody may actually facilitate the growth of transplanted tumors by the phenomenon of enhancement discovered by Kaliss.

The third era, which overlaps the second, began with the discovery that animals can be made resistant to transplants of a tumor from an isogenic donor in which the tumor originated and can develop antibodies active *in vitro* against it. In other words, it has now been shown that tumors may have specific antigens which are absent from the normal tissues of the host. (See Old and Boyse (37).) This conclusion was put on a firm basis by Prehn and Main (38), and subsequent work by Prehn has shown that many chemically induced mouse tumors are capable of immunizing animals of the same in-bred strains, which can be explained only by the existence of tumor-specific antigens. Such antigens have also been demonstrated by similar methods in sarcomas induced by plastic films, in polyoma virus tumors, and in various leukemias, some known to be viral. It is also possible in some cases, as Prehn and the Kleins have shown, to make an animal resistant to its own autochthonous tumor. This suggests the possibility that some tumors may be antigenic in the individual in which they originate and, therefore, possibly to some extent under immunologic control. Such control may explain some of the phenomena of tumor restraint discussed earlier. On the other hand, the most obvious feature of many malignant tumors is that they are accepted by the host without an active reaction of the type provoked by bacterial or other foreign invaders and thus continue to grow and kill the host. Work is in progress in several laboratories on possible exploitation of tumor antigenicity in the control of cancer, but this does not appear to have met with much success and is beyond the scope of the present discussion.

Specific tumor antigens have not been demonstrated in all tumors, and there is evidence that some tumors lack certain antigens present in the normal tissue of origin (47). An important role for loss of tissue-specific antigens has been postulated by H. N. Green (26) in his immunologic theory of cancer, but this has not been proved.

Burnet (7) has recently followed up earlier suggestions that tumors may be regarded from some points of view as homografts. Thomas (46) and others suggested that homograft immunity is based biologically on the need to control abnormal cells arising in the body by somatic mutation or some equivalent process. Burnet points out that the transfer of tissues from one animal to another is unknown in nature, and that homograft immunity is only one example of a general process that he calls "immunological surveillance." According to these ideas, tumors are clones of malignant cells, possibly arising by somatic mutations and possibly having altered surface properties which make them liable to immunologic surveillance. The carcinogenic process will be successful only if this immunologic control can be overcome.

It seems probable that immunologic factors do play a role in controlling the growth of tumors and possibly in their induction, but the importance of these factors is still unknown.

## NUTRITIONAL FACTORS IN CARCINOGENESIS (44, 45, 50)

Peyton Rous (39), as with so many other aspects of cancer research, was among the first to consider the influence of diet on carcinogenesis. More recently, Albert Tannenbaum has been a major contributor to this field, and the following discussion is drawn largely from his work and from that of F. R. White.

The results of much animal investigation and the available clinical data in man indicate that the genesis of tumors is enhanced by adequate nutrition. On the other hand, caloric restriction markedly inhibits the development of tumors, and proportions of dietary fat, protein, and vitamins below the minimum level for good nutrition also tend to repress tumor formation. An exception to this occurs in the liver, where carcinogenesis may be augmented by dietary deficiencies.

The effect of nutritional factors on the growth of tumors is less definite, and it is well known that tumors may increase in size while the animal is losing weight. Although deprivation of calories, proteins, or other dietary components may retard the growth of tumors, the deleterious effect on the host may outweigh any advantage gained. According to Tannenbaum, the diet must be regarded as a modifier, rather than an initiator, of the carcinogenic process.

It is obviously important to know the extent to which a particular tumor can be influenced by food intake in experiments on the effects of hormones, drugs, and other factors on tumor growth. Failure to consider the influence of nutritional factors may lead to erroneous interpretation of such experiments. Generally speaking, underfeeding increases the latent period, decreases the incidence of spontaneous neoplasms, and hinders the establishment and retards the growth rate of transplanted tumors.

## GENERAL CONCLUSIONS

From the preceding, very inadequate survey of some of the available information, it is clear that much remains to be learned about growth and trophic factors in carcinogenesis. One depressing but unavoidable conclusion is that, apart from the very rare cases of apparently spontaneous regression, all tumors are irreversible. This, unfortunately, seems also to apply to conditional or dependent tumors which sooner or later progress to autonomy. There appears to be no known way of permanently arresting the progression of cancers, but there are notable examples, especially in the field of hormonal control, in which the tumor can be checked for long periods.

If these conclusions are accepted, one implication relating to health hazards of man's environment is clear. Until much more information is available from fundamental research, efforts to prevent cancer are more likely to succeed than efforts to cure it. An obvious approach to the prevention of cancer is to reduce the exposure of the human population to carcinogenic stimuli in the environment and, where possible, to eliminate exposure altogether.

A large number of environmental carcinogens are known (28), and it is probable that more will be found. No attempt at a general discussion of this field can be made, but this presentation will be concluded by a very brief account, as an example, of some nitroso carcinogens and related compounds whose carcinogenic action has been discovered only relatively recently and which may be present in the environment (11, 30, 31).

This group of carcinogens has proved effective in all species in which tests have been reported. Virtually all organs are sensitive to one or another nitroso carcinogen. Several nitroso compounds have been shown to induce tumors in several organs after single doses in the rat. Some of these carcinogens, including those capable of inducing cancer after single doses, are rapidly eliminated from the body, and there is evidence that some react with the genetic material of somatic cells.

Nitrosamines are formed by reaction of secondary amines with nitrites, and there is a recorded example of the formation of lethal concentrations of the powerful carcinogen dimethylnitrosamine in fishmeal preserved with sodium nitrite and used for feeding sheep in Norway (Sakshaug, Sögnen, Hansen, and Koppang, 1965). The presence of nitrosamines has been very recently reported in tobacco smoke (43) and in white flour (32). Nitrate fertilizers are a potential source of nitrate, since nitrate reductase enzymes are found in plants and micro-organisms. Any secondary amines present in plants might therefore be nitrosated to yield the corresponding nitrosamine. Clearly, further research is required to determine whether nitrosamine can arise in plants in this way.

## REFERENCES

1. BEGG, R. W. (1958). Tumor-host relations. *Advan. Cancer Res.* **5**, 1–54.
2. BERENBLUM, I. (1954). Carcinogenesis and tumor pathogenesis. *Advan. Cancer Res.* **2**, 129–175.
3. BERENBLUM, I. (1964). The two-stage mechanism of carcinogenesis as an analytical tool. *In* "Cellular Control Mechanisms and Cancer" (P. Emmelot and O. Mühlbock, eds.), pp. 259–267. Elsevier, Amsterdam.
4. BERENBLUM, I., AND SHUBIK, P. (1947). A new quantitative approach to the study of the stages of chemical carcinogenesis in the mouse's skin. *Brit. J. Cancer* **1**, 383–391.
5. BIELSCHOWSKY, F., AND HORNING, E. S. (1958). Aspects of endocrine carcinogenesis. *Brit. Med. Bull.* **14**, 106–115.
6. BOYLAND, E. (1958). The biological examination of carcinogenic substances. *Brit. Med. Bull.* **14**, 93–98.
7. BURNET, F. M. (1964). Immunological factors in the process of carcinogenesis. *Brit. Med. Bull.* **20**, 154–158.
8. CLAYSON, D. B. (1962). "Chemical carcinogenesis," pp. 93–96. Churchill, London.
9. CLIFTON, K. H. (1959). Problems in experimental tumorigenesis of the pituitary gland, gonads, adrenal cortices, and mammary glands: a review. *Cancer Res.* **19**, 2–22.
10. DRUCKREY, H. (1959). Pharmacological approach to carcinogenesis. *In* "Ciba Foundation Symposium on Carcinogenesis," (G. E. W. Wolstenholme and M. O'Connor, eds.), pp. 110–130. Churchill, London.
11. DRUCKREY, H., PREUSSMANN, R., AND SCHMÄHL, D. (1963). Carcinogenicity and chemical structure of nitrosamines. *Acta Unio Intern. Contra Cancrum* **19**, 510–512.
12. EVERSON, T. C. (1964). Spontaneous regression of cancer. *Ann. N. Y. Acad. Sci.* **114**, 721–735.
13. EVERSON, T. C., AND COLE, W. H. (1956). Spontaneous regression of cancer: preliminary report. *Ann. Surg.* **144**, 366–383.
14. FOULDS, L. J. (1954). The experimental study of tumor progresison: a review. *Cancer Res.* **14**, 327–339.
15. FOULDS, L. J. (1961). Progression and carcinogenesis. *Acta Unio Intern. Contra Cancrum* **17**, 148–156.
16. FOULDS, L. J. (1958). The natural history of cancer. *J. Chronic Diseases* **8**, 2–37.
17. FOULDS, L. J. (1954). Hormones and cancer. *Practitioner* **192**, 370–375.
18. FRIEDEWALD, W. F., AND ROUS, P. (1944). The initiating and promoting elements in tumor production. *J. Exptl. Med.* **80**, 101–125.
19. FURTH, J. (1953). Conditioned and autonomous neoplasms: a review. *Cancer Res.* **13**, 477–492.
20. FURTH, J. (1954). The concept of conditioned and autonomous neoplasms. *In* "Ciba Foundation Symposium on Leukaemia Research," pp. 38–41. Churchill, London.
21. FURTH, J. (1957). Discussion of problems related to hormonal factors in initiating and maintaining tumour growth. *Cancer Res.* **17**, 454–463.

22. FURTH, J. (1959). A meeting of ways in cancer research: thoughts on the evolution and nature of neoplasms. *Cancer Res.* **19**, 241–258.
23. FURTH, J. (1961). Vistas in the etiology and pathogenesis of tumors. *Federation Proc.* **20**, 865–873.
24. FURTH, J. (1963). Influence of host factors on the growth of neoplastic cells. *Cancer Res.* **23**, 21–34.
25. FURTH, J., KIM, U., AND CLIFTON, K. H. (1960). On evolution of the neoplastic state: progression from dependence to autonomy. *Natl. Cancer Inst. Monograph* **2**, 148–177.
26. GREEN, H. N. (1958). Immunological basis of carcinogenesis. *Brit. Med. Bull.* **14**, 101–105.
27. GREENE, H. S. N. (1951). A conception of tumour autonomy based on transplantation studies: a review. *Cancer Res.* **11**, 899–903.
28. HUEPER, W. C., AND CONWAY, W. D. (1964). "Chemical carcinogenesis and cancers." Thomas, Springfield, Illinois.
29. KIDD, J. G. (1961). Does the host react against his own cancer cells? *Cancer Res.* **21**, 1170–1183.
30. MAGEE, P. N., AND BARNES, J. M. (1967). Carcinogenic nitroso compounds. *Advan. Cancer Res.* **10**, 164–246.
31. MAGEE, P. N., AND SCHOENTAL, R. (1964). Carcinogenesis by nitroso compounds. *Brit. Med. Bull.* **20**, 102–106.
32. MARQUARDT, P., AND HEDLER, L. (1966). Über das Vorkommen von Nitrosaminen in Weizenmehl. *Arzneimittel-Forsch.* **16**, 778–779.
33. MILGROM, F. (1961). A short review of immunologic investigations on cancer. *Cancer Res.* **21**, 862–868.
34. MÜHLBOCK, O. (1963). Hormones in the genesis of cancer. *Neoplasma (Bratisl.)* **10**, 337–342.
35. MÜHLBOCK, O., AND BOOT, L. M. (1959). The mechanism of hormonal carcinogenesis. *In* "Ciba Foundation Symposium on Carcinogenesis," (G. E. W. Wolstenholme and M. O'Connor, eds.) pp. 83–94. Churchill, London.
36. MÜHLBOCK, O., AND VAN NIE, R. (1961). Hormone dependence and autonomy. *In:* Biological approaches to cancer chemotherapy. (R. J. C. Harris, ed.) Academic Press, New York.
37. OLD, L. J., AND BOYSE, E. A. (1964). Immunology of experimental tumours. *Ann. Rev. Med.* **15**, 167–186.
38. PREHN, R. T., AND MAIN, J. M. (1957). Immunity to methylcholanthrene-induced sarcomas. *J. Natl. Cancer Inst.* **18**, 769–778.
39. ROUS, P. (1914). The influence of diet on transplanted and spontaneous tumours. *J. Exptl. Med.* **20**, 433–451.
40. ROUS, P., AND KIDD, J. G. (1941). Conditional neoplasms and subthreshold neoplastic states. *J. Exptl. Med.* **73**, 365–390.
41. SALAMAN, M. H. (1958). Cocarcinogenesis. *Brit. Med. Bull.* **14**, 116–120.
42. SALAMAN, M. H., AND ROE, F. J. C. (1964). Cocarcinogenesis. *Brit. Med. Bull.* **20**, 139–144.
43. SERFONTEIN, W. J., AND HURTER, P. (1966). Nitrosamines as environmental carcinogens, II: Evidence for the presence of nitrosamines in tobacco smoke condensate. *Cancer Res.* **26**, 575–579.
44. TANNENBAUM, A. (1958). Nutrition and cancer. *In* "Physiopathology of Cancer." (F. Homburger, ed.) pp. 517–562. Cassell, London.
45. TANNENBAUM, A., AND SILVERSTONE, H. (1953). Nutrition in relation to cancer. *Advan. Cancer Res.* **1**, 451–501.
46. THOMAS, L. (1959) Discussion of paper by P. B. Medawar on Reactions to homologous tissue antigens in relation to hypersensitivity. *In* "Cellular and Humoral Aspects of the Hypersensitive States," (H. S. Lawrence, ed.) pp. 529–532. Cassell, London.
47. WEILER, E. (1959). Loss of specific cell antigen in relation to carcinogenesis. *In* "Ciba

Foundation Symposium on Carcinogenesis," (G. E. W. Wolstenholme and M. O'Connor, eds.) pp. 165–178. Churchill, London.
48. WEISBURGER, J. H. (1964). On mechanisms of liver carcinogenesis: effect of pituitary hormones and N-hydroxy-2-fluorenylacetamide. In "Cellular Control Mechanisms and Cancer" (P. Emmelot and O. Mühlbock, eds.) pp. 300–306. Elsevier, Amsterdam.
49. WEISBURGER, J. H., PAI, S. R., AND YAMAMOTO, R. S. (1964). Pituitary hormones and liver carcinogenesis with N-hydroxy-N-2-fluorenylacetamide. *J. Natl. Cancer Inst.* **32**, 881–904.
50. WHITE, F. R. (1961). The relationship between underfeeding and tumor formation, transplantation and growth in rats and mice. *Cancer Res.* **21**, 281–290.
51. WOODRUFF, M. F. A. (1964). Immunological aspects of cancer. *Lancet* **2**, 265–270.

## Commentary

### EMMANUEL FARBER

*Department of Pathology, University of Pittsburgh,
Pittsburgh, Pennsylvania*[1]

It would be premature to conclude that all carcinogenic processes start with a single irreversible phase setting in motion an inevitable progression of events.

Epstein and Merkow in our laboratory have shown that the liver responds to all hepatic carcinogens with the production of localized proliferations of paler cells. With acetylaminofluorine and ethionine large hyperplastic nodules containing these cells are particularly common and provide material for study. They contain large amounts of glycogen, much of which is retained when the animal is fasted. They also show a progressive decrease of glucose-6-phosphatase activity. So far attempts at transplantation have been unsuccessful.

Why should all the different carcinogens, which produce a wide spectrum of final malignancy effects, produce this common early response associated with altered glycogen regulation? This response may provide an opportunity for studying the question as to whether there is a single, irreversible, initiating step in carcinogenesis.

[1] Present address: Fels Research Institute, Temple University School of Medicine, Philadelphia, Pennsylvania.

# The Mechanism of Some Structural Alterations of the Lung Caused by Environmental Stresses[1]

PAUL GROSS

*Department of Occupational Health, Graduate School of Public Health, University of Pittsburgh, Pittsburgh, Pennsylvania*

The basic structure of the lung, the alveolar wall, is composed of two embryologically different tissues, the capillary network, which is derived from the mesoderm; and the alveolar membrane, an entodermal derivative. Each of these components reacts in a specific manner to inhaled environmental irritants.

In general, the alveolar capillary responds to an irritant that is sufficiently aggressive to penetrate the protective alveolar membrane. This response consists of pulmonary edema or of acute pneumonia or both.

The alveolar membrane responds also to irritants that are less aggressive. Structural alterations produced by irritation of the alveolar membrane consists of thickening of alveolar walls by proliferation of alveolar cells and the elaboration of a precollagenous argyrophilic supporting stroma. This proliferated tissue is avascular and may extend into air spaces to a degree that may result in their occlusion. The proliferative reactions of the alveolar membrane are reversible as long as the stroma remains precollagenous. In general, this reactive stroma (of ectodermal origin) collagenizes not at all or over a period of months and years compared to the stroma of mesodermal origin which collagenizes regularly in two to three weeks.

Both components of the alveolar wall may react to an inhaled irritant, and the structural alterations specific to each component may be present.

The pulmonary clearance mechanism is the most important factor that determines if a reaction to injury from an inhaled irritant will occur and where. Multifocal alveolar membrane reactions are more common than diffuse reactions and occur in regions where the clearance mechanism was incapable of removing the irritant. Conversely, a diffuse alveolar proliferative reaction to an inhaled irritant generally implies a diffuse inadequacy of the clearance mechanism relative to the irritant. The hypersensitivity type of reaction, which may also be diffuse, is an exception to this rule.

---

[1] Summary of presentation, taken from its publication in *Arch. Environ. Health* 14, 883–891 (1967).

## Commentary

Hollis G. Boren

*University of Colorado Medical Center, Denver, Colorado*

The balloon and tube model of the lung traditional with the physiologist, and the membrane model favored by the biochemist, can be considerably improved in the light of recent knowledge, but there is still too little known about the structure and function of other than the ciliated cells of the airways. Newer techniques of lung preservation, such as the vascular-perfusion-fixation procedure, permit better examination of detailed anatomical and histological relationships. Electron microscopy permits more exact identification of cell types.

The alveolar wall consists of two types of cells resting on a basement membrane. The Type A cell is an attenuated epithelial cell, with a few mitochondria, a few lysosomes, and a little endoplasmic reticulum. The Type B cell is larger and foamy looking, and may extend through the alveolar wall. It forms tight junctions with neighboring cells, has microvilli, many vesicles, many lysosomes, many mitochondria, many ribosomes, and a well-developed rough endoplasmic reticulum. Within the alveolar wall there exists real tissue space with regular components of collagen and microfibers. It is not known whether lymphatic clefts or nerve elements are present. On the surface of the alveolar walls is found a third type of cell, without a basement membrane, the alveolar macrophage. It is large, with pseudopods, phagocytic vacuoles, many cytoplasmic inclusions, many mitochondria, some giant mitochondria, and large numbers of lysosomes.

Gross' paper raised the following questions: (1) When does cellular proliferation become irreversible? (2) When does arborescence of the argentophile fibers become irreversible? (3) Can clearance mechanisms be blocked or overburdened to give damage? (4) Are healthy lungs free of microorganisms? (5) What cells maintain and repair lung structure? (6) What cells convey information that the lung has previously contacted an environmental agent? (7) What is the mechanism of the "biologic activity" of fine dust that leads to fibrosis?

# Mechanism of Bronchial Response to Inhalants[1]

ARTHUR B. DUBOIS[2]

*Graduate Department of Physiology, The School of Medicine,
University of Pennsylvania, Philadelphia, Pennsylvania*

Inhalants capable of producing functional bronchial responses appear to constitute one of the potential health hazards in man's environment. I shall attempt to discuss the mechanisms of functional bronchial response to inhalants.

The scope of this presentation does not include the mechanisms of sensitization to allergenic substances, or mechanisms of inflammatory or neoplastic change in the tissues. However, be it understood that the functional response may be accompanied by a tissue response which is more persistent than the functional response, either because the clearance of the inhaled agent from the site of action is incomplete or because the tissues have been altered at the same time that a reversible functional change occurred.

Although some inhalants which result in changes in lung function leave a normal person healthy, similar exposures may kill other persons, those unable to tolerate an attack of bronchoconstriction, bronchial inflammation, or increased bronchial secretions. And even under controlled laboratory conditions and with normal subjects, one has to cope with an occasional severe attack of bronchoconstriction resulting from the inhalation of an agent to which the average response is rather mild.

### CURRENT KNOWLEDGE AND REVIEW OF SOME OF THE LITERATURE

Inhalation of irritant substances produces attacks of (a) cough, (b) tachypnea, (c) substernal burning sensation, (d) wheezing, (e) dyspnea, and (f) tightness of the chest. The last three symptoms are associated with bronchoconstriction. Such attacks sometimes are followed by acute bronchitis and bronchopneumonia.

Filley, Hawley, and Wright (14) described the bronchoconstrictor effect of colloidal silica in isolated, perfused guinea pig lungs. Banister, Fegler, and Hebb (3) found bronchoconstriction in rabbits, cats, and dogs after the inhalation of ammonia or phosgene. The bronchoconstriction which occurred during the first minute of exposure was abolished by cutting the vagi. However, the bronchoconstriction which was present between the first and fifth minutes after exposure was not abolished by cutting the vagi. The bronchoconstriction was potentiated by eserine, an effect reversed by atropine. The authors concluded that phosgene and ammonia may act directly on peripheral structures, in addition to having an action mediated through the central nervous system and carried by peripheral nervous structures and the release of acetyl-

---

[1] The work reported here was supported by Public Health Service Grant H4797.
[2] Career Investigator, National Institutes of Health.

choline. Furthermore, the bronchoconstriction was elicited in isolated, perfused lungs.

Dautrebande et al. (7), found that the inhalation of small particles caused an alteration in the rate and depth of breathing and in the resting respiratory level of lung volume in man. Sim and Pattle (38) exposed human subjects to concentrated individual constituents of smog and found that the subjects developed coughing, rales, and bronchoconstriction. After repeatedly exposing themselves to sulfuric acid particles, both authors felt that they had acquired a persistent bronchitis, and had developed increased susceptibility to further inhalations of bronchial irritants.

Dautrebande and I (12) measured the airway resistance, lung compliance, and functional residual capacity in man before and after exposure to inert dust particles. The average airway resistance increased from 1.46 cm $H_2O$ per liter per second before dust to 2.63 cm $H_2O$ per liter per second after dust. The amount of dust delivered in the inspired air was about 10 mg of coal dust or 5 mg of calcium carbonate powder for each 2-min period of inhalation. Aluminum oxide powder and fine carbon particles made from India ink aerosol also produced the effect. The bronchoconstriction was reversed or prevented by inhalation of a bronchodilator aerosol. Inhalation of an aerosol generated from 0.5% to 2% concentrations of carbachol solution produced similar increases of airway resistance.

Further measurements of airway resistance after dust inhalation by normal persons and by patients who had lower airway obstruction were carried out by Constantine et al. (6) and Dautrebande et al. (9) Also, McDermott (22) exposed humans to coal dust concentrations of 8, 9, 19, 33, and 50 $mg/m^3$ for 4-hr periods. No changes of airway resistance were found at the 8 or 9 $mg/m^3$ dust concentration levels. But after 19 $mg/m^3$, airway resistance was found to have increased. After 33 and 50 $mg/m^3$, airway resistance increased further, and this increase was correlated with the weight of the coarse particles (3.6–7 $\mu$ in diameter). Difficult breathing was encountered at 33 and 50 $mg/m^3$. Lloyd and Wright (20) found consistent increases of airway resistance after dust particles.

A similar effect on airway resistance was found as a result of inhaling cigarette smoke, and was attributed to the particles therein (21, 26). The unevenness of the distribution of ventilation and blood-flow ratios after dust inhalation was described by Raine and Bishop (32) and Norris and Bishop (30).

Now let us consider the effects of gases, rather than particles. Amdur and Mead (1, 2) reported a useful method for studying the effects of gases on unanesthetized guinea pigs' lungs. Frank et al. (15) found that, in man, sulfur dioxide concentrations of 1 ppm caused no significant change in pulmonary resistance, but 5 ppm increased the pulmonary resistance by 39%, and 13 ppm increased it by 72%. Similarly, Nadel et al. (27, 28) found that sulfur dioxide, 4–6 ppm, decreased the ratio of airway conductance to thoracic gas volume by 39%; atropine prevented this. In cats, the increase of pulmonary resistance from sulfur dioxide was prevented by cold block of the cervical vagosympathetic nerves, or by atropine given intravenously.

Let us turn to the more recent literature on mechanisms of bronchoconstriction in response to inhalants. The anatomy of the nerves and nerve endings in the lungs was described by Larsell and associates (16–19). The bronchial walls are well supplied with sensory nerve endings down to the level of the alveolar atria. Parasympathetic ganglia are found deeper within the bronchial walls. Widdicombe (39, 40) investigated the neurophysiologic function of the receptors in the airways of cats and their response to sulfur dioxide and ammonia. He then joined Nadel and his associates (29, 42) in studies of the afferent and efferent impulses resulting from tactile stimulation or of dust inhalation into the airways. This work, and subsequent work of Nadel et al. (24, 27, 28), emphasized the reflex nature of such bronchoconstriction, mediated through afferent sensory fibers running with the vagus nerve, and efferent vagal pathways. Nadel, Colebatch, and Olsen (25) also described a different mechanism, which is peripheral airway constriction triggered by substances reaching the lungs via the bloodstream, and not mediated through a vagal reflex. These concepts have been further elaborated (4, 5, 11, 24, 31). A recent chapter by Widdicombe (41) reviewed the factors which modify airway tone.

Although reflex bronchoconstriction mediated via vagal afferent and efferent fibers must be an important mechanism in bronchoconstriction after the inhalation of small amounts of dust or of sulfur dioxide and other gases, it cannot be the sole mechanism, because mechanical alterations of the lungs have been found in isolated, perfused lungs (8, 14) and after vagal section (3) or after lungs exposed to dust were removed from the body (10, 33–36). At least part of the alteration of the pressure-volume curve of the lungs after dust may be due to a change of surface tension (23, 37) and part to release of histamine or acetylcholine (35).

## STATE OF THE ART

The episode of bronchoconstriction (increased airway resistance) which one finds in man after inhalation of a few breaths of certain substances (histamine aerosol, carbachol aerosol, inert particles, cigarette smoke, or sulfur dioxide gas) has its onset within a minute of the inhalation, reaches a maximum in 5–10 min, and then decays to normal over the next 45 min. Re-exposure at any stage reinduces a similar increment of airway resistance. The effects are reversed or prevented by administration of such bronchodilators as isoproterenol aerosol.

The degree of response depends on the amount of substances administered, probably following an S-shaped dose-response curve. There is variation of response among individuals, and in the same individual on different occasions. But the effect of inert dust particles occurs regularly and is easily demonstrated, provided the methods used to measure the effect are appropriate. Measurements of airway resistance by means of a body plethysmograph, or pulmonary resistance by the esophageal pressure method, are most suitable. The forced expiratory volume curve shows very little change, and it is therefore a poor method to use for such experiments.

The phenomenon of bronchoconstriction after dust or sulfur dioxide exposure as analyzed in animals appears to have a strong reflex component in which the receptors are similar to the touch, pain, or cough receptors in the trachea and bronchi; the afferent nerves accompany the vagus, and the efferent pathways are vagal, resulting in constriction of the smooth muscle in the trachea and bronchi. So far as we know, the cerebral cortex does not enter into this reflex arc, and therefore the response probably cannot be conditioned psychologically.

The fact that excised lungs show a response to inhalation of dust particles must mean that additional local mechanisms come into play. The release of histamine and acetylcholine and the action of dust on surface tension forces of the lungs all have been postulated.

In addition, many ganglionic cells are demonstrable in the lungs, and it is not known whether local reflexes play a part in the bronchoconstriction after dust exposure.

## SPECULATIONS CONCERNING IMPLICATIONS

The defenses which man and other mammals possess against injurious gases in the inspired air include brief periods of breathholding, as by closure of the larynx; a conscious person either voluntarily shuts off the source of the contaminant or moves away to breathe fresh air somewhere else. His nose usually tells him when it is time to take such action.

But the nose is not remarkably sensitive to particles suspended in the inspired air, and cannot detect foreign bodies aspirated through the mouth and into the trachea. Laryngeal closure and the cough reflex provide a defense against these objects. But fine particles usually enter the bronchial tree unannounced.

Once in the bronchi, irritant gases and fine particles trigger certain responses consisting of cough, bronchoconstriction, a burning sensation behind the sternum, tachypnea, and increased mucus production.

But this response is a double-edged sword. The bronchoconstriction may suffocate an individual whose pulmonary function is already barely adequate, and the added mucus, mucosal swelling, or cellular proliferation in the airways may close air passages if they are already narrowed. Furthermore, toxic gas inhaled in sufficient concentration may produce necrosis of the bronchial epithelium, inhibition of ciliary action, and increased susceptibility to respiratory infection.

If the exposure is a single event, and if the person survives the episode, then healing processes occur, and recovery may be complete or almost complete. But in repeated small injuries, the reflex response or healing process itself may become overexaggerated, and the individual may be disabled more by his defense mechanism than by the inhalant.

Clearly, there are responses which occur entirely within the body. But there are also defenses which depend on an individual's outside reaction, such as avoidance or flight; and the healing process may be augmented by outside assistance in the form of medical treatment. Besides this, in a group wherein persons interact with each other, an individual adversely affected warns others of the danger. His nasal sensations, substernal pain, cough, wheeze, and dyspnea say: "Beware!"

Technology has given us amplification of the sensory mechanisms which warn the individual of danger. That is, we can measure gas and particle concentrations in the inspired air at levels lower than the olfactory threshold. We can use physiologic instruments to measure effects on the bronchial tree at levels below the threshold at which the physiologic response is perceived by the person so affected.

Natural philosophy has given us the tools of rationalization required to record, systematize, and generalize the experiences of individuals reacting to inhalants into a pattern of cause-and-effect relationships such that the resulting model has predictive value; we can forsee, within statistical limits, what will happen to an individual as a result of exposing him to certain levels of certain agents in the inspired air.

But technology and philosophy are ineffective unless they are capable of modifying the response. This implies the need for a combination of critical evaluation, decision, and effective action.

To summarize: Man's physiologic reaction to injurious inhaled agents can be amplified with methods for measurement of his response. The response, measured objectively, can be compared with the concentration and duration of the agent as amplified and measured objectively. Abstractions (models) can be developed relating the particular type of response to the inhaled agent. A review and appraisal of the meaning of these schemata in terms of their implications to individuals and to the society which they constitute can be undertaken.

The appropriate actions which could be taken theoretically include several alternative possibilities. First, one could sense agents in the inspired air and then limit their concentrations. Second, one could modify the individual's response to such agents by intervening in the physiologic pathways through which such responses occur. Third, one might adapt individuals to a new way of life in which their bodies and behavior became accustomed to new levels of exposure. Fourth, one could select groups of individuals who show natural adaptation, and use these groups to do special jobs or live or work in special places. Fifth, one could modify the human race in such a way that its responses differed from those which now exist.

Obviously, the most attractive of these alternative possibilities is the first. But it is also expensive. Thus, a comparison of the desirability of the various alternatives ultimately must be made. Some of the philosophy which forms a prelude to such a comparison has been described by Dubos (13). The present symposium would seem to be one which would lead to the elaboration of processes whereby man, in a rational society, can either control his own environment or else adapt himself and his race to those factors which he cannot control.

## REFERENCES

1. AMDUR, M. O. (1959). The physiological response of guinea pigs to atmospheric pollutants. *Intern. J. Air Pollut.* (*London*) 1, 170–183.
2. AMDUR, M. O., AND MEAD, J. (1956). The respiratory response of guinea pigs to inhalation of ethylene oxide. *Arch. Indust. Health* 14, 553–559.
3. BANISTER, J., FEGLER, G., AND HEBB, C. (1949). Initial respiratory responses to the intratracheal inhalation of phosgene or ammonia. *Quart. J. Exptl. Physiol.* 35, 233–250.
4. COLEBATCH, H. J. H., OLSEN, C. R., AND NADEL, J. A. (1966). Effects of 48/80 on the mechanical properties of the lungs. *J. Appl. Physiol.* 21, 379–382.
5. COLEBATCH, H. J. H., OLSEN, C. R., AND NADEL, J. A. (1966). Effect of histamine, serotonin, and acetylcholine on the peripheral airways. *J. Appl. Physiol.* 21, 217–226.
6. CONSTANTINE, H., DAUTREBANDE, L., KALTREIDER, N., LOVEJOY, F. W., JR., MORROW, P., AND PERKINS, P. (1959). Influence of carbachol and of fine dust aerosols upon the breathing mechanics and the lung volumes of normal subjects and of patients with chronic

respiratory disease before and after administering sympathomimetic aerosols. *Arch. Int. Pharmacodyn.* **123,** 239–252.
7. DAUTREBANDE, L., ALFORD, W. C., HIGHMAN, B., DOWNING, R., AND WEAVER, F. L. (1948). Studies on aerosols. V. Effect of dust and pneumodilating aerosols on lung volume and type of respiration in man. *J. Appl. Physiol.* **1,** 339–349.
8. DAUTREBANDE, L., DELAUNOIS, A. L., AND HEYMANS, C. (1958). New studies on aerosols. IV. Effect of fine dust particles on excised guinea pig's lung before and after sympathomimetic aerosols. *Arch. Int. Pharmacodyn.* **116,** 187–208.
9. DAUTREBANDE, L., LOVEJOY, F. W., JR., AND CONSTANTINE, H. (1960). New studies on aerosols. XI. Comparative study of some methods used for determining constriction and dilation of the airways after administering pharmacological or dust aerosols. Sensitivity of the plethysmographic method. *Arch. Int. Pharmacodyn.* **129,** 469–491.
10. DAUTREBANDE, L., ROBILLARD, E., ALARIE, Y., AND PAGANUZZI, P. (1963). New studies on aerosols. XIX. Influence of air-borne submicronic, inert particulates administered to guinea pigs prior to establishment of acute, total lung atelectasis immediately followed on the excised lungs by determination *in vitro* of "pressure volume" hysteresis curves. *Arch. Int. Pharmacodyn.* **144,** 278–292.
11. DEKOCK, M. A., NADEL, J. A., ZWI, S., COLEBATCH, H. J. H., AND OLSEN, C. R. (1966). New method for perfusing bronchial arteries: Histamine bronchoconstriction and apnea. *J. Appl. Physiol.* **21,** 185–194.
12. DUBOIS, A. B., AND DAUTREBANDE, L. (1958). Acute effects of breathing inert dust particles and of carbachol aerosol on the mechanical characteristics of the lungs in men. Changes in response after inhaling sympathomimetic aerosols. *J. Clin. Invest.* **37,** 1746–1755.
13. DUBOS, R. "Man Adapting." (1965). Yale Univ. Press, New Haven, Connecticut.
14. FILLEY, G. F., HAWLEY, J. G., AND WRIGHT, G. W. (1945). Toxic properties of silica: Bronchoconstrictor effect of colloidal silica in isolated perfused guinea pig lungs. *J. Ind. Hygiene* **27,** 37–46.
15. FRANK, N. R., AMDUR, M. O., WORCESTER, J., AND WHITTENBERGER, J. L. (1962). Effects of acute controlled exposure to $SO_2$ on respiratory mechanics in healthy adults. *J. Appl. Physiol.* **17,** 252–258.
16. LARSELL, O. (1921). Nerve terminations in the lung of the rabbit. *J. Comp. Neur.* **33,** 105–131.
17. LARSELL, O. (1923). The ganglia, plexuses and nerve terminations of the mammalian lung and pleura pulmonalis. *J. Comp. Neur.* **35,** 97–132.
18. LARSELL, O., AND BURGET, G. E. (1924). The effects of mechanical and chemical stimulation of the tracheobronchial mucous membrane. *Am. J. Physiol.* **70,** 311–321.
19. LARSELL, O., AND MASON, M. L. (1921). Experimental degeneration of the vagus nerve and its relation to the nerve terminations in the lung of the rabbit. *J. Comp. Neur.* **33,** 509–516.
20. LLOYD, T. C., JR., AND WRIGHT, G. W. (1963). Evaluation of methods used in detecting changes of airway resistance in man. *Am. Rev. Respirat. Dis.* **87,** 529–537.
21. LOVEJOY, F. W., JR., AND DAUTREBANDE, L. (1963). New studies on aerosols. XX. Effects of cigarette smoke on the airway conductance in smokers and non-smokers. *Arch. Int. Pharmacodyn.* **143,** 258–267.
22. MCDERMOTT, M. (1962). Acute respiratory effects of the inhalation of coal dust particles. *J. Physiol. (London)* **162,** 53P.
23. MILLER, D., AND BONDURANT, S. (1962). Effects of cigarette smoke on the surface characteristics of lung extracts. *Am. Rev. Respirat. Dis.* **85,** 692–696.
24. NADEL, J. A. (1965). Structure-function relationships in the airways: Bronchoconstriction mediated via vagus nerves or bronchial arteries; peripheral lung constriction mediated via pulmonary arteries. *Med. Thorac.* **22,** 231–243.
25. NADEL, J. A., COLEBATCH, H. J. H., AND OLSEN, C. R. (1964). Location and mechanism of airway constriction after barium sulfate microembolism. *J. Appl. Physiol.* **19,** 387–394.

26. NADEL, J. A., AND COMROE, J. H. (1961). Acute effects of inhalation of cigarette smoke on airway conductance. *J. Appl. Physiol.* **16**, 713–716.
27. NADEL, J. A., SALEM, H., TAMPLIN, B., AND TOKIWE, Y. (1965). Mechanism of bronchoconstriction during inhalation of sulfur dioxide. *J. Appl. Physiol.* **20**, 164–167.
28. NADEL, J. A., TAMPLIN, B., AND TOKIWE, Y. (1965). Mechanism of bronchoconstriction during inhalation of sulfur dioxide; reflex involving vagus nerves. *Arch. Environ. Health* **10**, 175–178.
29. NADEL, J. A., AND WIDDICOMBE, J. G. (1962). Reflex effects of upper airway irritation on total lung resistance and blood pressure. *J. Appl. Physiol.* **17**, 861–865.
30. NORRIS, R. M., AND BISHOP, J. M. (1966). The effect of calcium carbonate dust on ventilation and respiratory gas exchange in normal subjects and in patients with asthma and chronic bronchitis. *Clin. Sci.* **30**, 103–115.
31. OLSEN, C. R., COLEBATCH, H. J. H., MEBEL, P. E., NADEL, J. A., AND STAUB, N. C. (1965). Motor control of pulmonary airways studied by nerve stimulation. *J. Appl. Physiol.* **20**, 202–208.
32. RAINE, J. M., AND BISHOP, J. M. (1964). Alterations in ventilation and perfusion relationships in the lung after breathing inert dust on bronchoconstrictor aerosols and after short periods of voluntary hyperventilation. *J. Clin. Invest.* **43**, 557–570.
33. ROBILLARD, E., AND ALARIE, Y. (1963). Mechanical properties and histamine and serotonin content of guinea pig lungs as influenced by microparticle inhalation. *Can. J. Biochem. Physiol.* **41**, 2177–2182.
34. ROBILLARD, E., AND ALARIE, Y. (1963). Pressure-volume curves in isolated atelectatic rat lungs after aluminum oxide microparticle inhalation. *Can. J. Biochem.* **41**, 1257–1265.
35. ROBILLARD, E., AND ALARIE, Y. (1963). Static volume-pressure relations in guinea pig lungs after inhalation of aluminum or iron oxide particles and bronchodilator aerosols. *Can. J. Biochem.* **41**, 461–468.
36. ROBILLARD, E., ALARIE, Y., PAGANUZZI, P., AND DAUTREBANDE, L. (1964). New studies on aerosols. XXI. "Pressure-volume" curves obtained on isolated atelectatic rats' lung after short inhalations of various insoluble submicronic and submicroscopic particles. *Arch. Int. Pharmacodyn.* **147**, 220–228.
37. ROSENBERG, E., ALARIE, Y., AND ROBILLARD, E. (1962). Effect of dust and aerosol inhalation on surface and tissue elasticity of rat lungs. *Can. J. Biochem.* **40**, 1359–1365.
38. SIM, V. M., AND PATTLE, R. E. (1957). Effect of possible smog irritants on human subjects. *J. Am. Med. Assoc.* **165**, 1908–1913.
39. WIDDICOMBE, J. G. (1961). Action potentials in vagal efferent nerve fibres to the lungs of the cat. *Arch. Exptl. Pathol. Pharmacol.* **241**, 415–432.
40. WIDDICOMBE, J. G. (1954). Receptors in the trachea and bronchi of the cat. *J. Physiol. (London)* **123**, 71–104.
41. WIDDICOMBE, J. G. (1966). The regulation of bronchial calibre. *In* "Advances in Respiratory Physiology" (C. G. Caro, ed.), pp. 48–82. Williams & Wilkins, Baltimore, Maryland.
42. WIDDICOMBE, J. G., KENT, D. C., AND NADEL, J. A. (1962). Mechanism of bronchoconstriction during inhalation of dust. *J. Appl. Physiol.* **17**, 613–616.

## Commentary

HOLLIS G. BOREN

*University of Colorado Medical Center,
Denver, Colorado*

The following questions are raised by DuBois' paper: (1) What is the stimulation site of agents causing bronchial constriction? (2) How do such agents penetrate the mucus of airways or the surface lining of alveoli? (3) Does chronic bronchial constriction result in irreversible changes? (4) Can studies be devised to show that surfactant is continuous with the mucus layer?

Exposure of animals to maximum concentrations of an environmental agent makes extrapolation to naturally occurring human exposures difficult. A system in which a random background of exposures prevailed might be more realistic.

Future work should emphasize chronic studies, sequence studies, effect of synergists, immune and adaptive phenomena, cell culture responses, and proliferative responses of newborn or fetal animals.

Most important is the realization that the lung is not an inert membrane for the passive transfer of gases, nor a lipid membrane upon which environmental agents impinge, but a living tissue having cellular and subcellular components whose functional capabilities are largely unexplored.

# Principles and General Concepts of Adaptation[1]

C. LADD PROSSER

*Department of Physiology & Biophysics,
The University of Illinois, Urbana, Illinois*

My assignment in this conference is to present some of the principles of physiological adaptation from the viewpoint of a general biologist. First, I wish to present some definitions in the hope that this conference may agree on a common terminology. It is unfortunate that in many standard works; for example, the Handbook of Physiology (3, 5, 15, 18), certain words are used by different contributors with very different meanings. I hope to show that the general terminology developed by comparative physiologists is applicable to human physiology. Further, while most of my examples will be drawn from temperature stresses, the terminology is equally applicable to such other environmental parameters as hypoxia, osmotic and ionic stresses, and nutritional factors.

The word "adaptation" has many meanings and should be defined in each context in which it is used. If I were to ask 100 biologists for the meaning of adaptation, I might get 100 different definitions. It may well be that this word is no longer precise enough for serious usage. For some, adaptation refers to those properties (usually anatomic) which have been selected over long periods of evolution to permit survival. For others, adaptation refers to the rapid decay of an excitatory process, as in sensory adaptation. For still others, adaptation refers to any homeostatic reaction. In the present discussion, adaptation refers to any property of an organism which permits physiological activity and survival in a specific environment; adaptation is characteristically related to stressful components of the environment although it may relate equally well to a total environment. Adaptive characters have genetic basis but may be expressed according to environmental needs.

A response is a direct reaction, either adaptive or nonadaptive to an environmental stimulus. It is usually, but not necessarily, reversible. An example would be a direct change in rate of a chemical reaction with temperature, as in a $Q_{10}$ effect.

Adaptive variations may be measured in individuals, populations, or higher taxonomic categories. They include anatomic, physiologic, and biochemical characteristics of individual organisms which relate these individuals adaptively to a specific environment.

In an evolutionary sense, only those variations that are adaptive are retained. Natural selection is the only known mechanism for fixation of adaptive variations and forms the basis for speciation.

[1] Research on which this paper is based was aided by NSF Grant GB4005.

A physiological concept of biological species may be derived as follows: If no two species can occupy the same ecological niche or the same geographic range throughout their life cycles, it follows that every species must be uniquely adapted to its particular niche and range. Hence, if we could quantitatively describe the physiological adaptedness of a species to its ecological niche and geographic range, we would have a truly meaningful description of the species. One of the goals of environmental physiology is to achieve some understanding of the molecular basis for natural selection.

However in the context of this conference adaptive variations are considered for individuals, not species.

"Homeostasis" is a term which has been broadened from the original meaning of Cannon to refer to self-stabilizing states or organisms, societies, and computers. Physiologically, we can use "homeostasis" not only for maintenance of constancy of the internal environment, but also for maintenance of constancy of energetics, of work capacity, of self-identity, and of independence of the external environment. The essence of living things is to be at once apart from their environment and a part of the environment. Homeostasis describes the sum total of properties contributing to the "apartness" of individuals.

## GOALS OF ENVIRONMENTAL PHYSIOLOGY

The goals of environmental physiology may be listed sequentially in three categories:

1. To describe adaptive variations in organisms—whether individuals, populations, races, or species. Much remains to be done in this descriptive phase with respect to human adaptive variation. It is not always easy to identify how a given character is adaptive—for example, pigmented skin in Negroes. Biologic meaning of some characters may be found only in genetic linkage and pleiotropism, as in sickle-cell anemia.

2. To discover the origin of such variations as are described—whether genetic or environmentally induced. Genetically determined variation is based on selection from random variation and is transmitted in the genotype. Environmentally induced, or nongenetic, variations may be transitory for individuals or may be transmitted culturally from generation to generation.

Genetically determined variation can generally be distinguished from nongenetic by acclimation or acclimatization supplemented by breeding experiments. Acclimation consists of those conpensatory changes which occur in individuals under controlled laboratory conditions where only one environmental parameter is varied, for example temperature or photoperiod. Acclimatization consists of those adaptive changes which occur under natural conditions when multiple factors vary, as in different climates, seasons, and geographic conditions. For example, the changes occurring in a rodent out-of-doors in the winter differ from the changes in similar animals in a cold room. Identification of critical environmental stressors is not always clear when they are multiple—temperature, photoperiod, nutrition, and others.

Acclimation in man is difficult to obtain, although acclimatization has been much studied. Prolonged exposures in decompression chambers have provided a basis for comparing acclimation to low pressure with acclimatization in high mountains. Acclimation to heat or cold has been studied with periodic, repetitive, or 1-week exposures in temperature chambers, but how this differs from continuous exposure for long times at a constant temperature is not established. Life in the Arctic or Antarctic is very different from that in a cold room. Evidence is now accumulating that animals from very uniform environments may lose their capacity for acclimation while those which live in highly variable environments have maximal capacity for acclimation. This has practical importance for the adaptability of transplanted populations and species.

The genotype sets the limits within which environmentally induced variation can occur, that is, the range of phenotypic variability. A major task in human biology is to separate the two causes of variation-genetic and environmentally induced.

3. To explain adaptive variation at all levels of organization. This includes a description of feedback interaction between environment and organism—for example, the regulation of synthesis of specific proteins by environmental stressors. Explanation of variation in biochemical terms may elucidate the molecular basis for natural selection.

## CRITERIA OF PHYSIOLOGIC VARIATION

Criteria of physiologic variation may be grouped into four broad categories (18).

1. Internal state for a given parameter as a function of that parameter in the environment. This is the essence of homeostasis and may occur in either of two patterns or a mixture of these:

*Conformity of internal state to the environmental state.* Familiar examples are poikilothermic and poikilosmotic animals in which temperature or osmotic concentration in body fluids and tissues rises and falls with these parameters in the external environment. Conformity is homeostatic in its broader meaning, in that enzyme systems of conformers can function over a wide range of internal environments. In man and other naked mammals, and in aquatic mammals, the skin temperature can vary widely according to environmental temperature, and enzymes of skin may be adapted to be active over a much wider temperature range than corresponding enzymes of liver.

*Regulation or constancy of internal state.* As the environmental parameter changes, feedback control reactions tend to maintain internal constancy; at some limiting gradient, these control reactions fail and the internal state changes abruptly. Familiar examples are maintenance of constant body temperature, constant blood sugar, and constant sodium-potassium ratios in body fluids. The cost of adaptation by internal regulation is considerable, but internal constancy permits body functions over a wide environmental variation.

In general, conformers tolerate wider internal variation, and regulators, wider external variation, with respect to a given parameter. Many animals show combinations of the two patterns,. regulation of one parameter or in one environmental range and conformity in another, or regulation at one stage in development or time in a life cycle. A hibernating mammal is a temperature conformer over a certain range when torpid, a regulator to relatively internal temperature when active.

A related measure of variation is the rate of recovery of internal state after deviation due to a stressing experience. For example, the pattern of recovery of normal water load after dehydration is specific to a species.

2. Rate functions. Any biologic function that can be measured as a rate with respect to time may be used as a criterion of adaptive variation when tested in different environmental states.

The time course of change in a rate function after an environmental change varies according to the magnitude of the stress, the rate of application of the stress, and whether the organism is a conformer or a regulator for the environmental parameter changed. In general, three phases can be recognized—an initial overshoot or undershoot in the rate function, a stabilized rate, and a period of compensation or acclimation. For example, when environmental temperature is suddenly raised, most poikilotherms show an initial rapid overshoot of such rate functions as metabolism; when temperature is lowered, they show an undershoot. Overshooting and undershooting reactions usually last only a few minutes. These reactions may be initiated reflexly by sensory stimulation, and their magnitude varies with amount of temperature change; initial overshoots and undershoots have been observed also in microorganisms. Similar initial overshoots of metabolism have been noted in man after sudden heat-stress application.

After the initial response in intact organisms, or immediately in enzymes and isolated tissues, a stable rate is achieved. This is the rate used for $Q_{10}$ measurements with temperature. The change in rate may be considered as a direct response to environmental change and may persist for many hours. If the animal is then returned to the original environmental state—e.g., the original temperature—the rate returns directly to the level prior to the deviation.

If the animal is held in the altered environment for days or weeks, some compensatory changes occur in rate functions. These compensations represent the processes of acclimation and if the animal is now returned to its original environmental condition, its rate function goes beyond the original level, thus indicating that a basic change in state had occurred during acclimation.

Two kinds of adaptive acclimation can be identified: capacity and resistance adaptations. The same two types of adaptation can be identified genetically. Capacity adaptation, whether acclimatory or genetic, are those permitting relatively normal rates of reaction in a mid-range of environmental variation. For poikilotherms, two schemes of capacity adaptations have been formulated. Precht's classification (17) for acclimation to lowered temperature is as follows (Fig. 1): no acclimation, i.e., the rate remaining at its initial stable level (type 4); perfect adaptation, i.e., the compensated rate at the low temperature being the same as at the higher temperature (type 2); partial acclimation with the rate between no and perfect compensation (type 3); overcompensation, i.e., the compensated rate in the cold higher than at the initial temperature (type 1); and inverse compensation, with the acclimated rate lower than

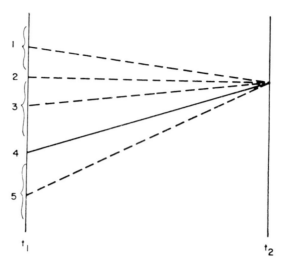

Fig. 1. Diagram representing patterns of acclimation of a rate function in a poikilotherm to cold. $t_2$, original temperature; $t_1$, lower temperature to which animal is transferred. Patterns: (1) overcompensation, (2) perfect compensation, (3) partial compensation, (4) no compensation, (5) converse or paradoxic acclimation. (From Precht (17)).

the stabilized rate, often explained by complicating secondary environmental factors (type 5). A second classification of rate acclimations is that of Prosser (19) (Fig. 2); this scheme considers not just two points, but the entire curve relating rates and temperature. The acclimated rate curve may be shifted by

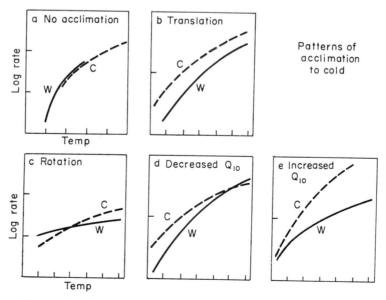

Fig. 2. Patterns of acclimation of rate functions in poikilotherms. Rate functions plotted logarithmically at different temperatures of measurement. W, warm-acclimated animals; C, cold-acclimated animals. (Modified from Prosser and Brown (19)).

translation to the left or right; it may be rotated with a change in temperature coefficient and with the intersection of the two curves at low, medium, or high temperatures; or the acclimation may consist of a combination of translation and rotation.

The pattern of acclimation of enzyme rates may differ according to the enzyme, the tissue in which it is found, and modifying factors. One tissue may show acclimation with respect to cold when tested *in vitro;* another tissue may show no such adaptation. The effect of a given environmental agent, such as cold or heat, may be directly on the cells of the adapting tissues; in other instances it may be mediated by endocrines or by central nervous influences. Acclimation may involve use of alternate metabolic pathways—for example, emphasis on the hexose monophosphate shunt in cold-acclimated fish.

3. Survival limits. At environmental extremes, adaptive variations can be measured in terms of survival of animals, developmental limits, and protein denaturation. Changes in limits of various rate functions with acclimation have been termed "resistance adaptations" by Precht (17). The limits of activity of given enzymes may change in the direction of compensation as they usually do in capacity adaptation, or they may not show any relation to capacity adaptation. Thus, the molecular mechanisms of resistance adaptation may be different from those of capacity adaptation. In nature, the distribution of organisms is limited more by environmental extremes than by environmental means.

4. Behavioral adaptations. The first line of defense against environmental stress in animals is usually reflex. In mammals, the reflexes of piloerection and of vasoconstriction in cold and of sweating, panting, and salivation in heat are well known. Longer-term behavioral acclimations are seen in nest-building, huddling in cold, in seeking underground burrows in heat. Many poikilotherms and poikilosmotic animals show "preferred" temperatures and salinities. In general, nervous or behavioral adaptations precede metabolic or biochemical ones. We found (20, 21) that in fish the temperatures for cold block of spinal reflexes, swimming, and conditioned reflexes could be modified by acclimation. The time for these nervous adaptations is significantly shorter that that for metabolic acclimation (2 days, compared with 10–14 days, in goldfish). It is probable that much of the individual variation described in this symposium results from conditioning of the nervous system, often at an early age. The nervous mechanisms of adaptation have been studied inadequately.

## MOLECULAR MECHANISMS OF ADAPTATION

The molecular bases for the criteria of adaptive variation may be enumerated for both genetically determined and environmentally induced variations as shown below.

### Primary Structure of Proteins; Conformational Changes

Genetic variations are manifested principally in the structure of proteins synthesized according to the genetic template. Changes in amino acid sequences are strictly fixed by the genotype and cannot be readily altered (except at nonphysiologic extremes). Amino acid sequences have, in a few instances, been correlated with adaptive significance. For example, in fishes and in certain worms, the melting or transition temperature of collagen is clearly correlated with the sum of proline and hydroxyproline content (12). In thermophilic bacteria, the protein flagellin is endowed with heat resistance because of an abundance of amino acids which provide high levels of hydrogen bonding (14).

## Generalized Synthesis of Protein and RNA

The translation of rate-temperature curves implies quantitative adaptive changes in enzymes. Measurement of turnover of protein by incorporation of radioactive amino acids (e.g., leucine $^{14}$C) shows that, after cold acclimation, fish have enhanced synthesis of protein (2). The increased synthesis is found in proteins of various subcellular fractions. Such generalized biochemical change is reasonable in view of the necessity for maintained balance in metabolism at different temperatures.

## Isozymes

Many proteins appear to exist in several forms, each with its own genetic basis. Differences in proportion of specific isozymes in different tissues are known—e.g., the lactic dehydrogenases (LDH's) in heart and skeletal muscle. Similarly, in compensatory temperature acclimation, Hochachka (6) has found that, in fish, certain LDH's are selectively formed in certain temperature ranges. It is probable that selective synthesis of specific forms of given proteins in different tissues under the stress of different environmental parameters (direct or indirect via hormones and nervous system) may be the most general mechanism of compensatory acclimation. The adaptive meaning of particular configurations is not yet understood. Such differential synthesis could account for rotational effects on rate-temperature curves.

## Lipids

It is well established that, in microorganisms as well as in animals, the lipids which are laid down at low temperatures tend to be more unsaturated and have longer chains than those deposited at high temperatures. Differences in lipids must reflect correlated differences in the enzymes of lipid synthesis. Differences are found in both neutral fats and phospholipids. However, the extent to which some enzymatic differences may be explained by differences in lipids of active intracellular membranes is not known. It is probable that some of the adaptations of central nervous systems are the result of alterations in membrane lipoproteins.

## Ions and Coenzymes

Changes in ion balance are known to occur during temperature acclimation in some poikilotherms. Whether these alterations are primary or are the result of other adaptive changes is not clear. If is of interest that, even in osmotic conformity, as in some marine molluscs, changes in total osmoconcentration seem to be mediated more by organic molecules than by inorganic ions.

## Enhancement of Certain Metabolic Pathways

Intermediary metabolism is replete with parallel pathways, each with several coupled enzymes. In the acclimated state, one path may be enhanced and another depressed, whether by synthesis or cofactors. For example in some poikilotherms in the cold, the hexosemonophosphate shunt and lipogenesis are enhanced relative to glycolysis, similarly in rats.

### APPLICATION OF THE PRINCIPLES TO MAN

Environmental physiology is a complex subject, and principles established for one organism or even for one enzyme in relation to homeostatic control may not apply to another system. However, similarities among organisms do exist and man must be considered as an animal with cultural modifications. The sequence of acclimatory modifications in any animal is relatively unrelated to the degree of regulation and conformity of internal state.

1. Genetic and environmentally induced differences among individuals and among "races" can be distinguished with difficulty. The work of Hammel (5), Edholm and Lewis (3), Wyndham et al. (27, 29), and Adams and Covino (1) indicates racial differences in respect to temperature regulation. The Bantu sweat rate is less than that of Europeans at the same elevated temperature; the critical ambient temperature for increasing metabolic heat production is lower for Negroes than for Caucasians; the Aborigines at night on the Australian desert tolerate a reduced body temperature, while Norwegians acclimated to cold ambient temperature had higher metabolism than when not acclimated.

Examples of environmentally induced variations are more familiar. In heat acclimation, the skin temperature for initiating of sweating is reduced. Vasomotor conditioning and reduced sensitivity of skin receptors occur in the hands of fishermen working with cold nets (3). A series of adaptive changes occur in persons living at high altitudes. In attempts to understand individual variations, the distinction between ontogenetic (developmental) and adults conditioning needs to be examined in greater detail.

2. The sequence of adaptive responses to environmental variation may be similar in mammals to that in lower animals, even though the manifestations are different. The general sequence of initial responses and late acclimation to cold in mammals is: pilomotor and vasomotor reflexes, which provide insulation; behavioral reactions, which are protective; shivering as a means of heat production; nonshivering thermogenesis, or heat production largely from nonmuscular tissues; long-term insulative changes, such as increased coat of hair; decreased sensitivity of skin receptors in some species; and elevated metabolic rate. Thus, in both homeotherms and poikilotherms, oxygen consumption increases after prolonged cooling. In poikilotherms, the net effect of compensatory acclimation of metabolic enzymes is to maintain relatively constant energy output at different temperatures (Fig. 3). Homeotherms show enhanced metabolism in the cold after they reach the stage of nonshivering thermogenesis; the net effect is to maintain constant body temperature with different energy outputs (Fig. 4). In poikilotherms and homeotherms, the enzymatic changes show

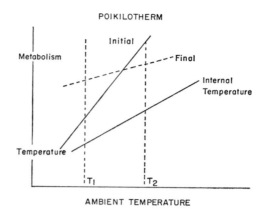

Fig. 3. Diagram of metabolism and body temperature of a poikilotherm which compensates metabolically by acclimation to different ambient temperatures. Internal temperature is same as ambient, initial metabolism shows high $Q_{10}$, acclimated final metabolism shows compensation at low ($T_1$) and high ($T_2$) temperature.

striking resemblances in metabolic compensations for cold, even though the purpose is different.

Compensation in the direction of reduced metabolism occurs in acclimation to heat in both poikilotherms and homeotherms. For example, laboratory rats show marked increase in $O_2$ consumption when transferred from ambient temperature of 28°C to 34°C, but after 2-4 days, their oxygen consumption returns to near its original value (11).

3. Acclimation of metabolism. The time after exposure to a given stressful environment and the rate of environmental change strongly influence the adaptive variations which are observed. An example is the metabolic compensation of homeotherms to prolonged cooling. For acute measurements of metabolism at different ambient temperatures, a thermoneutral zone is observed over which

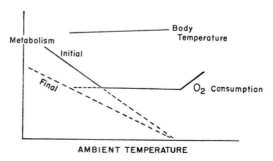

Fig. 4. Diagram of metabolism and body temperature of a homeotherm which maintains constant body temperature at different ambient temperatures. Metabolism is minimal over thermoneutral zone when insulative reactions adequately maintain body temperature. At lower temperatures, metabolism increases initially after acclimation (final), the critical temperature for metabolism increase may be lower and the rate of increase of metabolism per degree of lowering of ambient temperature may decrease. The slopes of metabolism—temperature curves may extrapolate to body temperature.

insulative responses are adequate to maintain constant body temperature. This thermoneutral zone may be broad, as in well-furred or feathered large animals, or it may be narrow, as in poorly insulated and small animals. At the critical low temperature, insulative mechanisms fail and heat production increases. Scholander et al. (23) found, and it has been frequently confirmed, that for moderate-sized mammals and birds the metabolism rises with reduced ambient temperatures according to Newton's law of cooling and that the metabolism-temperature curve tends to extrapolate to body temperature (Fig. 4). The critical temperature varies with degree of insulation, and the slope of the metabolism-ambient temperature curve may be modified by heat distribution in the animal tissues.

If, on the contrary, the observations are made chronically—that is, on animals acclimated to the temperatures at which measurements are made—very different results are obtained. For example, in grosbeaks the slope of the acclimated metabolism-temperature curve extrapolates not to 38°C, but to about 50°C. Also, the acclimated metabolism varies according to whether the cooling is cyclic or constant (25). Similarly, in cattle shifted from 65°F to 110°F over 6 hr the $O_2$ consumption rose abruptly as heat regulation was established, whereas when similar cattle were subjected gradually over a 2-month period to a comparable elevation of environmental temperature (from 53°F to 95°F), their metabolism actually decreased (13). The metabolism-ambient temperature curve for white-throated sparrows extrapolated to body temperature at night when insulation was good, but extrapolated to higher temperatures by day, when birds were active (7).

In man, the metabolism-temperature curve varies greatly according to the conditions of measurement and time for equilibration. Acutely measured curves may be curvilinear and may extrapolate to about 25°C or slightly higher. This must be related to heat transfer between body core and periphery. Also, the curves in man are much steeper when evaporative stress is high than in still air, and the curves in exercise are different from those obtained at rest. Measurements made in men, naked but exercising by bicycle to achieve comfort, extrapolate to 37°C (22). True acclimated curves have not been constructed at many temperatures for man, and present evidence fails to explain the marked difference between metabolism-ambient temperature curves for man and furred mammals.

## Uniqueness of Man

Despite the similarities among animals mentioned above, man shows so many differences from other animals that extrapolation is difficult. Some of the biologic characteristics which are prominent in man are the following:

1. Man constitutes one biologic species. There is no reproductive isolation among populations or races, except that which may be socially imposed. It is most unlikely that modern man can ever evolve into subspecies. This means that genetically all men are reproductively compatible. No other such cosmopolitan species exists.

2. Man shows remarkable phenotypic liability. Human populations are adapted

to a wide range of climates—dry and moist air, cold and hot climates, reduced oxygen at high altitudes. This phenotypic lability permits many environmentally induced adaptations; the genetic limits for these are extremely wide for man. As pointed out by Ernst Mayr (16), such phenotypic flexibility retards evolution and permits extension of range nongenetically.

3. Most of the adaptive characters of man are based on multiple alleles; these have been well analyzed for such proteins as hemoglobin (9). At the same time, pleiotropism is extreme; single cistrons lead to multiple effects. This is shown by correlated characters—for example, the relation between blood groups and certain types of carcinoma.

4. In human populations, balanced polymorphism is important. Individuals show great genetic diversity and in a population no two individuals (except for identical twins) are genetically identical. This means that certain recessive characters are bound to be maintained at a fixed ratio under steady-state conditions (as specified by the Hardy-Weinberg law), and the probability of eliminating such characters is virtually zero. Balanced polymorphism has been observed to impart certain advantages to populations, whether in nature or in breeding cages (4). First, it means that for most characters there is a predominance of heterozygotes with the resulting biologic advantages of heterosis. At the same time, some characters are carried in cryptic form. Heterozygotes tend to be more vigorous than homozygotes in general and to show greater success in variable and stressful environments, whereas homozygotes tend to be more influenced by the environment and heterozygotes more resistant. Balanced polymorphism can account for the maintained percentages of various characters in different populations. For example, most individuals homozygous for sickle-cell anemia die before the age of 5 years, yet heterozygotes are maintained at 40% in central Africa, 30% in India, and 17% in Greece. The correlation of heterozygosity for sickle-cell anemia with resistance to malaria is well known. Similarly, blood types are maintained in balanced proportions in populations.

5. Man has evolved culturally very rapidly, biologically very slowly. Cultural transmission provides for a wealth of adaptive traits, most of them behavioral. Culturally acquired traits—for example, in housing and clothing—permit survival in a wide range of climates. As a result, the microclimate in which man lives is remarkably uniform; an Eskimo in his furs lives at a microclimatic temperature similar to that of an Indian in the tropics. Cultural inheritance and cultural evolution are much less important in the adaptation of animals to environments, although nongenetically acquired behavior does exist.

In conclusion, it must be emphasized that the patterns and molecular mechanisms of both genetic and nongenetic adaptive variation in animals are worthy of study in their own right. This is part of our continuing effort to understand life processes, irrespective of direct application to man. The interactions between organisms and their environments are exceedingly complex, and no single limiting mechanism exists for one environmental parameter in a given species. The time course of acclimation and the different effects of varied rates of change of an environmental parameter must be carefully examined for all adaptive responses. There is need for much research on the mechanism of adaptive variation—studies at all levels of biologic organization. Extrapolation from animals to man must be viewed with extreme caution because of man's biologic uniqueness.

REFERENCES

1. ADAMS, T., AND COVINO, B. G. (1948). Racial variations to a standardized cold stress. *J. Appl. Physiol.* **12**, 9–12.
2. DAS, A., AND PROSSER, C. L. (1967). Biochemical changes in goldfish acclimated to high and low temperatures. II. Protein synthesis. *Comp. Biochem. Physiol.* **21**, 449–467.
3. EDHOLM, O. G., AND LEWIS, H. S. (1963). Terrestrial animals in cold: man in polar regions, In "Handbook of Physiology," Sect. 4. *Am. Physiol. Soc.* Washington, D. C. pp. 435–446.
4. FORD, E. B. (1965). "Genetic Polymorphism." M.I.T. Press, Cambridge, Massachusetts.
5. HAMMEL, H. T. (1963). Terrestrial animals in cold: recent studies of primitive man. In "Handbook of Physiology," Sect. 4, pp. 413–434. *Am. Physiol. Soc.*, Washington, D.C.
6. HOCHACHKA, P. W. (1966). Lactate dehydrogenases in poikilotherms: Definition of a complex enzyme system. *Comp. Bioch. Physiol.* **18**, 261–270.
7. HUDSON, J. W., AND KIMZEY, S. L. (1966). Temperature regulation and metabolic rhythms in populations of the house sparrow, *Passer domesticus*. *Comp. Biochem. Physiol.* **17**, 203–217.
8. IAMPIETRO, P. F., VAUGHAN, J. A., GOLDMAN, R. F., KREIDER, M. B., MASUCCI, F., AND BASS, D. E. (1960). Heat production from shivering. *J. Appl. Physiol.* **15**, 632–634.
9. INGRAHAM, V. M. (1963). "The Hemoglobins in Genetics and Evolution." Columbia Univ. Press, New York.
10. JANSKY, L. (1965). Adaptability of heat production mechanisms in homeotherms. *Acta Univ. Carolinae Biologica*, 1–91.
11. JOHNSON, H. (1967). Time course of oxygen consumption in rats during sudden exposure to high environmental temperature. *Life Sci.* **6**, 1221–1228.
12. JOSSE, J., AND HARRINGTON, W. F. (1964). Role of pyrrolidine residues in structure and stabilization of collagen. *J. Molec. Biol.* **9**, 269–287.
13. KIBLER, H. H. (1962). Oxygen consumption in cattle in relation to rate of increase in environmental temperature. *Nature* **186**, 972–973.
14. KOFFLER, H., MALLETT, G. E., AND ADYE, J. (1957). Molecular basis of biological stability to high temperature. *Proc. Natl. Acad. Sci.* **43**, 464–477.
15. LADELL, W. S. S. (1963). Terrestrial animals in humid heat: man, pp. 625–660. In "Handbook of Physiology," Sect. 4, *Am. Physiol. Soc.*, Washington, D. C.
16. MAYR, E. (1963). "Animal Speciation and Evolution." Harvard Univ. Press, Cambridge, Massachusetts.
17. PRECHT, H. (1958). Concepts of temperature adaptation of unchanging reaction systems of cold-blooded animals. In "Physiological Adaptation." (C. L. Prosser, ed.) pp. 50–78. *Am. Physiol. Soc.*, Washington, D. C.
18. PROSSER, C. L. (1963). Perspectives of adaptation: theoretical aspects. In "Handbook of Physiology," Sect. 4, pp. 11–25. *Am. Physiol. Soc.*, Washington, D. C.
19. PROSSER, C. L., AND BROWN, F. A. (1961). "Comparative Animal Physiology." Saunders, Philadelphia, Pennsylvania.
20. PROSSER, C. L., AND FARHI, E. (1965). Effects of temperature on conditioned reflexes and on nerve conduction in fish. *Zeit ch. vergl. Physiol.* **50**, 91–101.
21. ROOTS, B. J., AND PROSSER, C. L. (1962). Temperature acclimation and the nervous system in fish. *J. Exptl. Biol.* **39**, 617–629.
22. SCHOLANDER, P. F., ANDERSON, K. L., KROG, J., LORENTZEN, F. V., AND STEEN, J. (1957). Critical temperature in Lapps. *J. Appl. Physiol.* **10**, 231–234.
23. SCHOLANDER, P. F., WALTERS, V., HOCK, R., AND IRVING, L. (1960). Body insulation of some arctic and tropical mammals and birds. *Biol. Bull.* **99**, 225–236.

24. THOMPSON, G. R. (1962). Significance of haemoglobins S and C in Ghana. *Brit. Med. J.* **1**, 682–685.
25. WEST, G. C., AND HART, J. S. (1966). Metabolic responses of evening grosbeaks to constant and to fluctuating temperatures. *Physiol. Zool.* **39**, 171–184.
26. WYNDHAM, C. H. (1965). Adaptation of some of the different ethnic groups in southern Africa to heat, cold and exercise. *South African J. Sci.* **61**, 11–29.
27. WYNDHAM, C. H., MORRISON, J. F., WARD, J. S. BREDELL, C. A. G., VAN RAHDEN, M. E., HOLDSWORTH, L. D., WENZEL, H. G., AND MUNRO, A. (1964). Physiological reactions to cold of Bushmen, Bantu and Caucasian males. *J. Appl. Physiol.* **19**, 868–876.
28. WYNDHAM, C. H., PLOTKIN, R., AND MUNRO, A. (1964). Physiological reactions to cold of men in Antarctic. *J. Appl. Physiol.* **19**, 593–597.
29. WYNDHAM, C. H., WARD, J. S., STRYDOM, N. B., MORRISON, J. F., WILLIAMS, C. G., BREDELL, C. A. G., PETER, J., VAN RAHDEN, M. J. E., HOLDSWORTH, L. D., VAN GRAAN, C. H., VAN RENSBURG, A. J., AND MUNRO, A. (1964). Physiological reactions of Caucasian and Bantu males in acute exposure to cold. *J. Appl. Physiol.* **19**, 583–592.

# Human Genetic Adaptation[1]

## C. C. LI

*Graduate School of Public Health,
University of Pittsburgh, Pittsburgh, Pennsylvania*

Genetic changes in a population may be produced by gene mutation, migrations among population groups, or selection through differential reproduction of various genotypes, to name a few ways. The mutation is a "random" event not representing any particular adaptation to the environment. Mutations provide the ultimate source of new genes from which selections may be made, but do not by themselves explain adaptation to environment. Selection is the tool for genetic adaptation to environment.

The heritability of most common traits and diseases in man remains unknown. Heritability is not an intrinsic and fixed property, but varies with environmental conditions and relative gene frequencies. Three categories of traits usually have low heritability: those easily influenced by nongenetic (environmental) factors; those directly related with reproduction; and those long regulated by selection. The best approach to study of heritability in human populations is by the linear regression of child on parent, or the mid-parent if information is available on both parents. The study is plagued, however, by the influence of the home environment, by assortative mating, and by a certain amount of inbreeding.

In random matings, the influence of ancestral genes diminishes very rapidly with successive generations, whereas the social legacies may persist for many generations. The distribution of even a single gene pair rapidly gets complicated; for a trait controlled by many pairs of genes, any given class of parents can contribute to a wide variety of classes of offspring; and the converse is equally true.

Evolution is a series of step-by-step conditional developments, each generation providing the starting point from which development can take place, regardless of what happened in the past. Most selection models are limited to one pair of genes; even with computer help a general solution for two pairs is not yet possible. Virtually nothing is known about natural selection with respect to complex traits partly controlled by heredity and partly influenced by environment. It is virtually impossible to markedly decrease, let alone eliminate, rare recessives by selection.

Selection for one trait usually brings about correlated changes in other traits that were not considered in making the selection. The more severe the selection, the greater the appearance of correlated responses. Marked changes of character can be brought about by selective breeding of plants or animals, but only where

[1] Synopsis of presentation, which appears in full in "Essays in Evolution and Genetics in Honor of Theodosius Dobzhansky" (M. K. Hecht and W. C. Steere, eds.), pp. 545–577. Appleton, New York (1970).

the character has a value to the breeder irrespective of defects that might have been simultaneously produced. This is not possible where the total fitness of the product for the environment is a condition of survival.

Genetic deterioration in man is largely a bugaboo. The rate of accumulation of rare recessives when selection pressure is removed is extremely slow, and a new equilibrium value usually results which is not very different from the old. The case for a decline of intelligence through higher breeding rates in less intelligent groups is not only very unlikely, but unprovable. It is not possible in half a generation to detect changes which require 60 generations for significant development. The converse argument for active improvement of the human population through selective breeding is not only statistically improbable, but founders at the outset for lack of criteria of improvement. The chief advantages of having a large random mating population are its stability and its capacity to respond to selection through genetic diversity.

Taken as a whole, there is no environment that is inherently better than others, nor is there a genotype that is inherently better than others. The entire variation in fitness is due to interaction between the genotype and its environment. Since there exist many different genotypes in our population, it is important to provide different environments for the maximum welfare of all individuals.

## Commentary

BERNARD L. STREHLER

*Department of Biological Sciences, University of Southern California, Los Angeles, California*

Li's view seems to run counter to the classical one that the evolution of a higher form of life is due to the interaction of natural selection and of variation, and that part of this process must entail the elimination from the reproductive population of individuals that are less fit. Forces—no doubt subtle or rigorous selection—did operate to produce the amazing functional entities that we now are. It should be possible to eliminate an undesirable trait in one generation, if the heterozygotes could be identified and removed. The unpleasant environments in which our ancestors lived eliminated a great number of alternative forms of life.

# Adaptive Cycles[1]

JÜRGEN ASCHOFF

*Max-Planck-Institut für Verhaltensphysiologie,
Seewiesen und Erling-Andechs, Germany*

Of the numerous biological rhythms, four have evolved in response to geophysical cycles and may be termed "adaptive"; tidal, diurnal, lunar, and seasonal. During the course of evolution, the organism has developed autonomous periodic processes, the periods of which match those of the environmental process. The environmental periodicity no longer acts as the proximate cause for the biological rhythm, but only as a synchronizing agent (zeitgeber) for a self-sustained oscillation within the organism. Evidence for self-sustained rhythms is given by experiments in constant conditions. The 24-hour rhythm of activity and of rest continues in birds when illumination and temperature are held constant and food and water are continually available. The period, however, now deviates from strictly 24 hours—it becomes a "circadian" (about 1 day) rhythm which may "free-run" with its own frequency for months. There is also evidence for circatidal, circalunar, and circannual rhythms. The free running circadian rhythm is entrainable by physico-chemical as well as by social zeitgebers. For all animals studied so far, the light–dark cycle is the most powerful zeitgeber.

There is probably no function and no organ which does not show a circadian rhythmicity. Of special interest are rhythms in sensitivity. Mice are more likely to be killed by a given dose of ethanol when it is given at the beginning of the activity cycle; whole-body irradiation kills mice in about half the time when applied during the dark time; narcosis from pentobarbital lasts longer when induced at noon instead of at midnight. Halberg and co-workers, using the cosinor (least square spectrum) on a computer program, have determined the values for crest time and for amplitude of twelve functions and constituents in human blood. An organism is a different biochemical, physical, and physiological system at each hour of the day. A stimulus or stress will have different effects, depending upon the phase at which it encounters the circadian map. The temporal order represented by the phase map is not an irrelevant by-product of the basic circadian rhythmicity, but a necessary prerequisite for a healthy state.

Free-running rhythms in man can be demonstrated if the subject is kept in isolation and if all clues of true time are excluded. Under these conditions, a single free-running rhythm for all components tends to result, with the various components preserving their phase relationship. Dissociation can, however, occur. Activity and rest rhythms may drift away from the body temperature rhythm, for example. This may occur also where different zeitgebers are in competition

---

[1] Synopsis of presentation, which was published in *Intern. J. Biometeorol.* **10**, 305–324 (1966).

(e.g., illumination time and watch time), or where two zeitgebers preserve a twenty-four hour rhythm, but out of phase (e.g., social and work cycles in night-shift operators). Miners on night shift may "behave as nocturnal animals for potassium excretion, and as diurnal animals for the excretion of water" (M. C. Lobban).

There are several possibilities to characterize health hazards in man's environment. In view of the facts: (1) that our environment contains rigid temporal programs; and (2) that the organism has a temporal organization in accordance with these programs; stresses may be classified as follows: (a) continuous stress, (b) incidental stress, recurring at a random fashion, and (c) programmed stress.

Stresses of all three categories will interface with, and their effects will depend on, the rhythmic structure of biological processes. Some kind of stress follows a situation to which the organism has not been adapted so far and which may be out of the limits of its capability for physiological adaptation. However, what can be such a stress at one time will not be so at another time—because of the periodically changing responsiveness and adaptiveness of the organism. The existence of adaptive cycles within the organism means: the organism is programmed to do "the right thing at the right time," and to be prepared in advance for the circumstances which will ensue in the programmed environment. A stress, therefore, can be defined not only as an "abnormal" situation (at any time) but also as an otherwise "normal" situation at the wrong time. Along this line of thought, a "normal" environment can become a stress if the organism does the "right thing at the wrong time," or if it failed to do what the environmental program expects it to do.

In restricting the discussion to circadian rhythms as the best known examples for adaptive cycles, some concluding remark should be made.

1) Circadian rhythms are based on a multioscillatory system which probably has to be synchronized internally in order to keep the organism in a healthy state.
2) Internal synchronization seems to depend, at least partly, on the phase-setting effects of zeitgebers. This means: we need a periodic changing environment which provides powerful enough zeitegebers.
3) Neither the effectiveness of abiotic, biotic, and social factors as zeitgebers for man is well enough known, nor is the consequence of a longer state of external or internal desynchronization.
4) The organism is a different physico-chemical and psychological system at each hour of the day. The effect of a stress depends to a large extent on the phase at which it encounters the circadian system.
5) Effects of long lasting stresses may depend on the extent to which they interfere with the mechanism synchronizing the circadian system with the zeitgeber.
6) Any physiological characterization of stresses has to take into consideration the temporal organization of the organism as well as the environment.

# Environmental Factors in Aging and Mortality

BERNARD L. STREHLER[1]

*Aging Research Laboratory, Veterans Administration Hospital, Baltimore, Maryland 21218*

*Received February 6, 1967*

## I. BACKGROUND AND DEFINITIONS

Any detailed analysis of the effects of environmental factors on the tempo of aging in metazoa is hampered by two interrelated fundamental deficiencies in the present state of our knowledge of the biology of aging: (a) Although the general intrinsic sources of aging can be listed (Strehler, 1966c), the quantitative contribution that each potential cause makes is not known; (b) The fraction of the total decline in function that can be attributed to modifiable environmental factors is unknown.

Despite the dearth of information bearing on these key questions, certain data are available which have permitted some tentative conclusions. Moreover, this information suggests a challenging variety of both broad and specific research questions, whose answers should enable us to understand in reasonable completeness the interactions of genetics and environment which produce the gradual decline in function we define as aging.

Before a systematic consideration of the various facets of environmental-organismic interaction in aging is undertaken, a few basic definitions need to be stated. For the purposes of this discussion, *aging* is defined (Mildvan and Strehler, 1960; Strehler, 1962a) as those changes in structure and function that (a) occur, usually following the attainment of reproductive maturity; (b) that result in a decreased ability to do the work necessary to overcome environmental or internal challenges; and (c) that result in an increased probability of death with time. It is obvious that such deterioration in function can arise from two distinct and different primary sources: (a) a genetic program that produces senescence as a part of an overall program of development or (b) various random and unprogrammed events including the effects of harmful and modifiable environmental factors.

In this context, it is a curious fact that most modern theories of mortality ascribe the death of an individual at a given time to an unusually large environmentally generated challenge that operates against a decrease in functional resistance to such challenges (Sacher, 1956; Strehler and Mildvan, 1960). However, this decrease in function is largely believed to result from genetically programmed events. Thus, it is tacitly hypothesized by most students of gerontology that, despite any aging-accelerating influence the environment may impose, the

---

[1] Present address: Department of Biology, University of Southern California, and Rossmoor-Cortese Institute for the Study of Retirement and Aging, Los Angeles, California.

major decrement in state which occurs with time is built into the machine, not the environment.

Such an arbitrary separation of organism and milieu is obviously artificial. Nevertheless, the relative similarity of life tables for genetically similar species in grossly different environments and the great contrasts in life tables between genetically diverse species in "identical" environments, makes the generalization a plausible one. In view of the above, we have elsewhere suggested (Strehler *et al.*, 1959) that the complex of factors that produce the phenomenon of aging must also have the following characteristics: (a) Intrinsicality; (b) Universality (within a species); (c) Progressiveness; and (d) Deleteriousness. Interestingly, only the last named criterion, deleteriousness, distinguishes aging from normal developmental processes, a point we shall return to subsequently.

Whereas it is a simple matter to set up such a set of definitions and criteria their description in detailed terms and the assignment of origins clearly depends on a knowledge of underlying mechanisms. There are several ways in which these underlying mechanisms are being investigated, approaches which have been called the steamroller, the ram-rod, and the stilletto. The steamroller approach is based on the belief that an understanding of the origins of aging will ultimately arise from the complete understanding of all biological phenomena. Its proponents are enthusiastic about the accumulation of more and more information, relevant (to aging) a priori or not. The ram-rod approach is based on the assumption that a systematic and comprehensive description of the structural and functional properties of organisms of different ages will eventually yield the key missing clues to the origins of aging. The stilletto approach is based on the assumption that all possible deteriorative changes will occur at some rate, that related groups of such changes can be stated as a limited number of specific hypotheses, and that the economical experimental approach is to enumerate these hypotheses and then to test their quantitative contribution to the functional decline of senescence in real biological systems. Ideally, the design of experiments to test these hypotheses consists of the controlled manipulation of an environmental or genetic variable so as to produce an enhanced amount of change of the type postulated and then to observe the resultant effect on performance and/or age specific mortality rate.

In order to illustrate the principle and some details of the last named approach, the major general hypotheses of aging are enumerated in Appendix I, column A. Model experiments suitable for the evaluation of the hypotheses are shown in column B. In column C are listed some findings relevant thereto, and in column D is given the tentative conclusion.

A perusal of Appendix I makes it clear that the approach suited to systematic testing of hypotheses is also suited to the evaluation of contributions by an environmental agent to the deleterious changes that result in senescence, and that a by-product of such research might be the definition of an optimum environment, at least in terms of survival. The concept and definition of an optimum environment contains implicitly an a priori definition of optimum for what: survival? life expectancy? enjoyment? high population density? health? rapid evolution? It is only in terms of the arbitrary choice of the object of optimization that a given environment can be defined or conceived. In terms of its relevance to the aging process(es), the approach of an environment to an optimum is, in general, reflected in the type of survival curve given by a specific species. In very hostile milieus, the probability of death is high at all ages and no change in death rate may be observable versus age; in hospitable environments, animals which senesce show a gradual increase in the probability of death with age. This is illustrated in Fig. 1 in which the exponential (first order decay) survivorship curve given by many species in the wild is contrasted to the so-called rectangular curve shown by a species existing and aging in a very favorable environment. The demographic data presented in Fig. 1 can be depicted otherwise (see Fig. 2, curves A and D), and mortality rates in environments of intermediate hostilities are illustrated in Curves B and C.

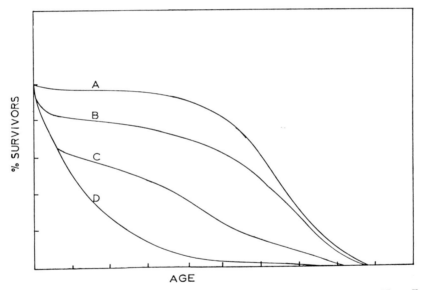

Fig. 1. Hypothetical survivorship curves. A, "rectangular" curve (after New Zealand 1934–1938) characteristic of hospitable environment. B and C, less hospitable environments (after Italy 1930–1932 [B] and India 1921–1930 [C]). D, survivorship in an extremely hostile environment in which the mortality rate is high (and constant) at all ages. Such a curve would be given by the age distribution of probability of death of victims of jet airplane crashes.

In another context, Strehler and Mildvan (1960) have developed and presented a mathematical model of mortality, a model which formalizes some of the relationships here depicted. The essence of this model is (a) that death occurs as a result of an organism's encountering a challenge that exceeds the maximum counteracting response of the challenged system; (b) that the frequency of such challenges decreases exponentially as the corresponding work demand increases linearly (as in the Maxwell-Boltzmann distribution of kinetic energy among gas molecules); and finally (c) that the well-known Gompertz equation ($R = R_0 e^{at}$), which describes the probability of death ($R$) at any age ($t$) ($R_0$ and $\alpha$ are appropriate constants), is generated by the interaction of this exponential distribution of challenges and the linear decrement in function (maximum work rate) that has been described for various physiological parameters (Strehler and Mildvan, 1960).

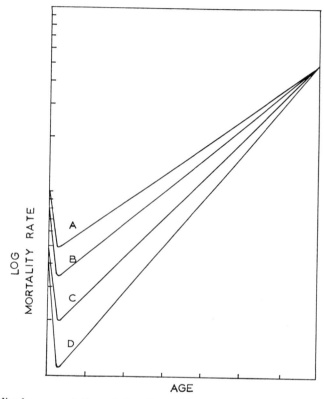

FIG. 2. Idealized representation of the Compertz plots (log probability of death vs. age) for environments of increasing hospitality (A to D). Note that the curves meet at advanced ages and that slope ($\alpha$) and intercept ($R_0$) are inversely correlated.

The algebraic treatment of this model has yielded certain predictions regarding $\alpha$ and $R_0$ which have been borne out in fact. One of these, illustrated in Fig. 3, shows the relationship of $\alpha$ and $R_0$ for a variety of different environments (data: human, U.N. Demographic Yearbook, 1955). The slope of the most probable line generated by the points depicted in this graph should, according to the theory, be equal to the reciprocal of the attrition coefficient (fraction of function lost/year), and the agreement between the value observed experimentally ($-0.007$ $-0.012$/year) and that calculated from these purely demographic data ($-0.0097$/year) is very good indeed. The absence of systematic curvature in the points plotted in Fig. 3 leads us to an important conclusion regarding the mechanism through which a poor environment shortens the mean length of life, and we shall now briefly consider this point.

Two general extreme positions might be taken regarding the effect of a "nonoptimal" environment on vital statistics. First, the effect of the environment may be predominantly to challenge the organism, without, however, causing lasting damage; alternatively, a harmful environment may produce permanent damage but produce little increase in the mortality rate at young ages. The consistancy of slope of the $\alpha$ versus $R_0$ plots for humans indicates that *the predominant difference* between the "good" and "bad" environments represented by these grossly different vital statistics is *in the level of challenge* and *not in the rate of damage*. In other words, the average rate of aging of human subpopulations is essentially unaffected by environmental factors; the differences in mean length of life are therefore ascribable to a greater frequency and magnitude of events that can cause death but not to an accelerated rate of decline in an unhospitable milieu.

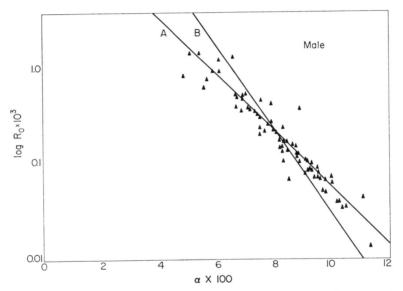

Fig. 3. Relation of $\alpha$ (Gompertz slope) and $R_0$ (extrapolated mortality rate at age = 0) for various male human national populations. Each point represents an individual country (data from U.N. Demographic Yearbook, 1955). Line A is a fit to the experimental data (Note absence of curvature in plot of "experimental points"). This indicates essential constancy of attrition rate ($\beta$) of reserve functional capacity, i.e., constant "rate of aging." The value of $\beta$ calculated from these data is about 1.4% (of vitality at age 30) loss per year. Real values are between 0.7 and 1.3% per year. Curve B gives line calculated for loss at rate of 1% per year.

This surprising and important conclusion (also consistent with the constancy of maximum human longevity since Roman times) has been confirmed in at least one group of experiments reported by Curtis (1963). He reported that repeated insults of a nonlethal nature (injection of bacterial toxins or of turpentine) were without effect on the life expectancy of survivors (see Fig. 4). Similarly, we have shown (Fig. 5) that thermal shocks are without permanent effect on the life expectancy of surviving *Drosophila*. Thus, neither of these acute types of challenges to life impairs the long-term functional capacity of the survivors.

On the other hand, it is well known that certain challenges do cause irreparable damage to cells and are reflected in permanent increases in the age-specific mortality rates of treated animals. Important among such stressful agents is exposure to ionizing radiation (Upton, 1962) (see Fig. 6) which will be considered more fully in a later section. Even infections (in animals or humans) do in some cases cause a significant increase in age-specific death rate. One need

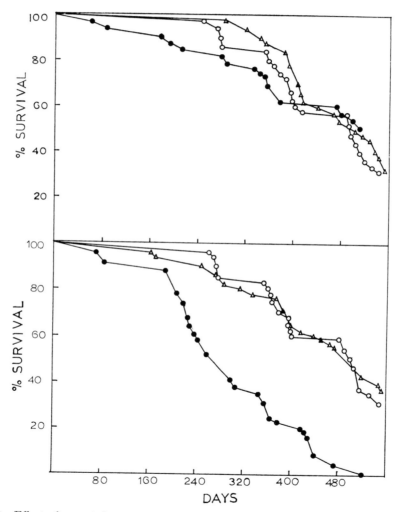

Fig. 4. Effect of repeated exposure to harmful substances on survivorship of mice. *Upper figure:* Mice exposed (three times a week) with large but nonlethal injections of tetanus toxoid (▲) and tetanus toxin (○) do not alter survivorship curves in mice. (●) Controls given saline. After Curtis (1963). *Lower figure:* Mice injected with turpentine (two times a month). (●) 0.03 ml; (△) 0.015 ml; (○) 0.5 ml saline. After Curtis (1963).

only recall the fact that certain latent types of mammary cancer are transmitted by a maternal milk factor (Smith, 1966) and that the life expectancy of victims of rheumatic fever, tuberculosis, polio, or nephritis is lower than that of suitable control groups. Ironically, some of the environmental factors promoting short-term survival (superabundance of food) may cause an increase in long-term debilities, e.g., atherosclerosis (Solomon, 1966).

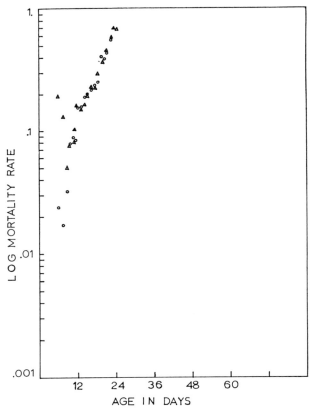

Fig. 5. Effect of exposure of *D. melanogaster* imagos to 38.5°C for one hour on the mortality rate of survivors. Note that exposure causes substantial mortality within 6–8 days of exposure but that the survivors beyond this initial period have the same life expectancy as the control flies. After Strehler (1961).

## II. CATEGORICAL ANALYSIS OF EXTRINSIC FACTORS AFFECTING THE RATE OF AGING

### A. Introduction

Any physical system in which each and every component is not replaced at a rate at least equal to its rate of deterioration or loss will undergo a gradual decline in state, that we may call aging. This applies to the obsolescence of automobiles and societies as well as the aging of skin, gut, blood, muscles, and nerves of human beings. Paradoxically, the aging of long-lived species is probably determined by very slow changes in very long-lived cell types such as muscle and nerve, rather than by changes in the steady state of the short-lived, regularly replenishing cell types.

In considering the environmental factors which are capable of producing the long-term decrements in function characteristic of aging, the primary question is, therefore, does the agent in question affect nonreplenishing structures, or does it affect systems which are regularly replaced. If it affects the former, it will probably produce long-term effects on function, i.e., it will accelerate aging.

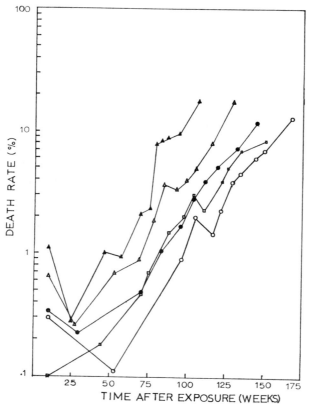

Fig. 6. Effect of exposure of LAF female mice to gamma irradiation on age specific mortality. (Mice were irradiated at 6–12 weeks of age.) After Upton (1962). (○) Non-irradiated controls (308 mice); (□) 223 rads (212 mice); (●) 368 rads (290 mice); (△) 578 rads (298 mice); (▲) 697 rads (266 mice).

To the extent that those varieties of harmful agents that were present within the ecological niche in which a species evolved, and that produce harmful effects before reproductive maturity is reached, will generally have favored the selection, evolutionarily, of those possible variants that possess resistance to the harmful factors, we may expect that most of the commonly encountered damaging agents will not result in permanent injury. Therefore, it is those environmental agents which are rarely encountered or to which adaptations cannot be evolved without concurrent or matching disability, whose effects form a residue that lasts for the life of the individual. We shall now proceed to consider systematically the major sources of such disruptive change.

B. *Sources of Disruptive Energy*

1. PHYSICAL SOURCES

The physical sources of potentially disruptive energy include all those processes and events which are capable of producing high local concentrations of energy—concentrations of sufficient magnitude to result in the cleavage of bonds, either individual bonds or groups of them. Such energy concentration may arise from several sources; of these, one of the most common is the absorption of radiant energy, which may occur in packages (i.e., quanta) of various sizes (Hollaender, 1954).

a. *Radiant energy.* The energy per quantum is directly proportional to the frequency. In the radio and infrared portion of the spectrum the energy per quantum is small compared with the energy required to disrupt a chemical bond. Therefore, if bonds are broken by the absorption of radiant energy of wavelengths longer than those visible to the eye, it is primarily through the summation of many absorption events that disruption may occur (Ferris, 1966). Such summation results in the elevation of the temperature of the absorber. The average kinetic energy of molecules at 27°C is about 600 cal/mole, whereas the energy of just-visible red light is about 30,000 cal/mole of quanta (einstein), or about 50 times higher than the average energy of a molecule at room temperature. To summarize: the absorption of quanta in the infrared and longer portions of the spectrum can produce damage, but only as a result of the elevation of the average temperature.

In the visible-and-shorter wavelength region of the spectrum, the energy per quantum is similar to or greater than the energy required to break the chemical bonds which maintain the intactness of molecules. Therefore, the exposure of molecules to this type of radiation may result in the cleavage of bonds and irreparable damage.

Damage as the result of bond cleavage by visible light is a rare event, because the energy involved (40–75 kcal) is somewhat less than the average energies of interatomic bonds (80–130 kcal). The process of photosynthesis (driven by visible light of about 42 kcal/einstein) may be looked upon as a special case in which the bonds involved cannot be completely cleaved by the energy available, but rather in which there is the formation of a weak bond from a strong one (e.g., $O-H \xrightarrow{h\nu} C-H$). In this case, light energy supplies the energy difference between the two states (Gaffron *et al.*, 1957; Rabinowitch, 1954; Strehler, 1952).

Although visible light is not, of itself, capable of breaking bonds, it may produce damage indirectly as a result of its photochemical catalysis of the formation of reactive species. An example is $XH_2 + O_2 \xrightarrow{h\nu} \cdot XH + HO_2$; $\cdot HO_2 + Y \rightarrow$ oxidation products of $Y$.

As we proceed from the visible region to shorter wavelengths of the spectrum, the energy/einstein is sufficient to break chemical bonds important to biological structure and function. Because of its limited penetration ability, damage by ultraviolet light occurs only in surface structures which can be exposed to it, e.g., the skin and its specialized appendages (Blum, 1959). It is significant that those gradual changes in the skin which are generally used as an index of chronological age (wrinkling, texture, etc.) occur primarily on the surfaces of the body exposed to light (Ma and Cowdry, 1950). The transformation of dermal connective tissue into an elastin-like component occurs in regions of the skin exposed to sunlight, but not in the protected regions (Hall et al., 1955). The damage is substantially less or absent in dark-skinned persons (Montagna, 1965), probably because melanin within the continually replenishing epidermal cells shields the underlying dermal structures from the main burden of ultraviolet radiation damage.

The efficiency of a given wavelength of light in producing damage is determined by two factors: (a) the number of quanta absorbed and (b) the way in which the absorbed energy is dissipated. The first factor is a function of the number of quanta arriving and of the probability that a given quantum will be captured by a sensitive molecule or structure. The second factor is affected by the nature of the absorber and by the environment in which the absorber finds itself. If the absorbed quantum is rapidly delocalized, i.e., if there is a small probability that the energy will be concentrated as kinetic energy in a given atom or electron, the molecule may dissipate its energy into the surroundings by a process known as internal conversion (Franck and Platzman, 1954), i.e., electronic excitation may result in the oscillation of atoms and thereby the energy may be transferred to the surrounding medium in small decrements. A second alternative mode of energy dissipation is the re-emission of light as fluorescence, light generally of lower energy (longer wavelength) than the absorbed quantum. The third alternative is the formation of a metastable state which may return to the ground state through stepwise energy release or through participation in a chemical reaction. Finally, the excited species may dissociate; that is, the electronic oscillation may make the structure unstable and it may decompose, either before or after the emission of an electron (ionization).

As the energy of the absorbed quantum is increased, the probability of ionization and dissociation tends to increase. For very short wavelength quanta, absorption may not even be required. Photons of gamma rays and X-rays, for example, acting as particulate projectiles rather than waves, produce the ionization or disruption of molecules they strike (Dictionary of Physics, 1961). This momentum transfer is a main source of the destructive effects of high-energy ionizing radiation on biological systems.

A few comments on the long-term effects of ionizing radiation are relevant. The damage caused by X-rays and other penetrating particles is relatively nonselective in the sense that any structure lying in the path of an oncoming photon may be disrupted by it (Ferris, 1966; Strehler, 1959). This applies to all cell components; however, damage to certain cellular structures will be more far-reaching in its effect than damage to other components. Thus, damage to structures which are present in very small numbers in a cell and which are also essential to its continued function is believed to be a primary cause of long-term radiation effects. The most obvious of such structures is nuclear DNA, for most genes are represented only twice in a diploid cell and damage to any one of them cannot easily be counteracted or repaired, particularly if both strands of DNA are destroyed. There is persuasive evidence that a DNA repair process occurs in cells (Curtis and Gebhard, 1958; Kaplan, 1966), but it is difficult to envision a mechanism of repair or replacement of an eliminated or totally mutilated segment of DNA if both strands of a given region are destroyed or altered. Because of its impartiality in producing damage and because the genetic apparatus is particularly vulnerable to its effects, ionizing radiation has been a widely used tool in certain types of aging research. It has been particularly useful in evaluating the somatic mutation hypothesis of aging (Alexander, 1966; Atwood and Pepper, 1961; Clark and Rubin, 1961; Curtis, 1966; Failla, 1960; Strehler, 1964a; Szilard, 1959).

The general consensus now appears to be that somatic mutation is *not* a major cause of the debilities of old age (Alexander, 1966). This conclusion is based primarily on the observation that dosages of radiation sufficient to induce many times the normally accumulated lifetime load of spontaneous mutations are without marked effect on the life-span of treated individuals (Curtis, 1966). Clearly, if mutations such as those produced by ionizing radiation are a key source of senescence, the doubling or tripling of the rate of their accumulation should double or triple the rate of aging. That it does not do so is a persuasive argument against the theory (Strehler, 1964a). For additional evidence bearing on the theory, see Appendix I.

b. *Kinetic energy transfer.* As an alternative to the absorption of radiant energy, the disruption of molecules and their organized aggregates may occur as a result of the transfer of kinetic energy from one component to another. This effect of momentum transfer may express itself at various levels of organization: individual molecule to whole organism.

Momentum transfer among molecules occurs as a result of collisions among them, and the rules governing these events are a part of classical physics. There is a feature of the distribution of energy among molecules which differentiates the phenomenon at this level from more gross effects and is responsible for the disruptive events that can occur, as it were, spontaneously, at ambient temperatures. This is the fact that, unlike heat distribution among large bodies (where heat flows from the warm to the cold object), the flow of molecular kinetic energy among molecules is determined by statistical rules governing multiple body collisions. A consequence is that some molecules will be at a higher temperature than other molecules (i.e., have more than the average kinetic energy) and some will be cooler than the average. This distribution of kinetic energy among molecules is described by the Maxwell-Boltzmann relationship (Moore, 1963) and is a direct consequence of the fact that collisions between elastic bodies with identical energies may result in a redistribution of energy in such a manner as to give some molecules more than the average energy and others less.

The Maxwell-Boltzmann relationship reveals that the number of molecules that possesses higher than average levels of energy decreases logarithmically as the energy they possess increases linearly. Very few molecules will therefore contain very large excesses of energy, but since such rare molecules can cause minute but severe damage to important structures, the long-term effects of such rare events may be considerable. Moreover, the disruptive effect may not be the direct consequence of structural disorganization, but rather of chemical reaction with components of the environment (such as $O_2$) potentiated by a local concentration of kinetic energy sufficient to overcome an activation energy barrier. Very slow oxidation reactions such as the yellowing of old newpapers, of varnish, and of lipid membranes are due to such events.

One of the most important destructive events occurring as a result of the accidental localization of heat energy in parts of living systems is the denaturation of proteins. This process, whose rate is characterized by a very high temperature dependence (high heat of activation), results in the precipitation of proteins and their inactivation as catalysts (McElroy, 1947).

Heat energy is one of the first studied of environmental factors that can affect the rate of senescence. A composite Arrhenius plot of temperature versus rate of aging for several species is illustrated in Fig. 7. The first quantitative study dealing with temperature effects on life span was reported by Loeb and Northrop (1917).

The effect of kinetic energy on structure is also evident in a variety of quite common and persistent effects on living systems. These effects, in contrast to the foregoing events, are due to the interactions of large aggregates of atoms. They constitute the cuts, bruises, abrasions, breaks, and all the physical injuries that make up the catalog of accidentally inflicted trauma.

Physical injuries in man's primitive environment were no doubt incurred by falls, battle wounds, attacks by wild animals, insect bites, etc., many of which are still with us today in a relaxed or exaggerated form. The common denominator of all such damage is that a force is applied locally to a tissue in such a manner that the cellular relationships are disrupted. If the agent of damage is a tooth, the mechanical advantage inherent in a wedge operates; if it is an automobile that is the destructive implement, the results may be more gross and permanent. In any event, there is no doubt that such injuries may result in permanent functional losses that may in turn prevent an individual from responding optimally to future challenges. But, by and large the contribution that a history of minor or even major physical injury makes to the debilities of the aged individuals in a population is probably small compared with the effects of more general deteriorative processes.

A third source of disruptive energy, of magnitude intermediate between events at the gross level and micro-accidents at the molecular level, are the effects of change in state. These are most marked in biological systems when ice crystals form as cells or tissues are frozen. The local forces generated within growing crystals of ice are sufficient to damage some cellular structures irretrievably (Meryman, 1956; Lovelock, 1954). Curiously, the formation of ice is a process which imparts a greater order into the water system. However, this achievement of greater order is accompanied by a somewhat increased volume, and this volume increment, combined with the penetrant qualities of newly formed ice crystals, and to local changes in concentration of dissolved substances, leads to greater disorder when the frozen system is thawed.

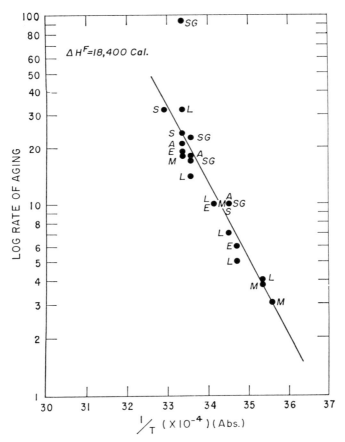

Fig. 7. Arrhenius plot of temperature vs. rate of aging for different invertebrate species. L, *D. melanogaster*; A, *D. melanogaster*; S, *D. melanogaster*; MS, *D. subobscura*; E, *Pinus tectus* Boie; M, *Daphnia magna*; SG, *D. melanogaster*, Gompertz plot. After Strehler (1962a).

## 2. Chemical Sources of Disruptive Changes

There are a variety of ways in which the occurrence of chemically reactive compounds within a living system may lead to the accumulation of damaged and/or altered structures. These include the following: (a) denaturation due to the local liberation of heat by highly exergonic (energy-liberating) reactions (Huennekens and Whiteley, 1960); (b) the incorporation of unreactive or inhibitory substances at important structural or catalytic sites of macromolecules (Harman, 1962; Harman and Piette, 1966); (c) the formation of intermolecular bonds (crosslinkages) by substances having more than one potentially reactive site (Bjorksten, 1958; Sinex, 1957; Verzar and Thoene, 1960); and (d) agents causing the dissolution of interatomic or intermolecular bonds (e.g., hydrolytic agents). We shall here briefly consider some representatives of these reactions and their implications for senescence.

a. *Denaturation reactions.* There are few published reports dealing with the contribution that denaturative processes make to aging. Denaturative reactions, because of their high activation energies, do not appear to be likely as central factors in the genesis of senescence for, from the studies of Loeb and Northrop and others, the temperature dependence of the aging process is more comparable to that of enzyme-catalyzed reactions (an activation energy of about 14–21 kcal) than to protein denaturation reactions (60 kcal activation energy).

In order to test the thesis that denaturation is an important component of the aging process, we have carried out experiments of the following nature (Strehler, 1961). Newly emerged *Drosophila* imagos were exposed for one hour to about 39°C. This critical exposure resulted in the death of about 50% of the flies within 2–5 days after exposure. A life table was constructed, based on the mortality experience of the survivors of this treatment regimen. A similar table for a control group of flies was also constructed. The results shown in Fig. 5 illustrate that the heat-shocked flies were not significantly "aged" by their brief exposures to unphysiologically high temperatures. Provided, as is likely, that this high temperature actually did produce substantial protein denaturation, we have concluded that neither protein denaturation, nor other processes with similar temperature dependences, are major contributors to the complex of events we define as aging. Experiments leading to substantially identical conclusions have been carried out by Maynard-Smith (1959). It would be of interest to determine whether persons frequently subjected to temperature at the upper extreme compatible with life (perhaps occupational hazards) (Strydom et al., 1963) show lasting damage as reflected in mortality statistics after they leave the environmental hyperthermic generators. Alternative human test material may be the mortality behavior of recovered victims of diseases that produce sustained or repeated high fevers, such as malaria.

If thermal denaturation per se is not verifiable at present as a central vector of cell and tissue aging in mammals, it should be pointed out that the structures affected by such denaturation may be substantially different from those damaged as a result of enzymically catalyzed chemical reactions, particularly those reactions characterized by large changes in free energy.

The enzymes that catalyze reactions resulting in a single-step liberation of large amounts of free energy are possibly particularly vulnerable to the local effects of the reactions they catalyze. This is apparent when we consider certain bioluminescent reactions. For example, luminous bacteria catalyze the emission of light with an energy/einstein of nearly 60 kcal (Harvey, 1940; Johnson, 1955; Strehler, 1955). If the energy potentially available for quantum ejection were channelled into the catalyst instead, it is more than probable that denaturation would be substantially accelerated over the rate of thermal distribution characteristic of the system (at the temperature at which the luminescent reaction occurs). Possibly, the stepwise release of energy in the mitochondrial sequence of reactions (Chance and Williams, 1956) is of value both in terms of the small and nondestructive packages of energy whose liberation it facilitates, as well as in terms of the conveniently utilizable size of the phosphoric anhydride free-energy content.

This latter type of denaturative process is substantially different from that which would be expected from random thermal accidents, and it is not possible (on the basis of current evidence) to rule out "catalyzed denaturation" of irreplacable structures as an important cause of senescence. Indeed, the primordial "rate of living" hypotheses of Rubner (1908) and Alpatov and Pearl (1929) are quite amenable to this model, for presumably the length of life would be determined by a total accumulated level of damage and this in turn would be dependent on the total metabolic flux occurring up to a given point in life. Even the observed temperature dependence of the rate of aging is amenable to the hypothesis, for the events leading to the denaturative liberation of energy would themselves possess ordinary enzymic activation energies.

b. *Binding reactions.* Several popularized theories of aging are based on the concept that reactive intermediates in metabolism or substances derived from the environment combine with enzymes or other functional structures of living systems and alter them in such a manner as to reduce their functional capacity. If the combining component is univalent, it might be called an occlusive inhibitor; if it is polyvalent it may act as a crosslinkage agent.

It should be pointed out that there are, unquestionably within biological systems, molecules capable of such reactions. What is not known is their amount, and more important, their cumulative effect on function. It is a potent (but all too often ignored) truism that all possible reactions occur at some rate. Biological systems selectively accelerate some of them (enzyme catalyses), but aging is a consequence of uncompensated reactions that occur at a rate too low appreciably to affect survival prior to the attainment of reproductive maturity.

In the following, we present certain experiments bearing on environmental factors that may modify the rate of aging or of age-correlated changes. The first of those agents is heavy water, and the experiments were designed to determine whether the incorporation of deuterium (in place of hydrogen) into a metazoan will substantially reduce life expectancy. The results obtained (see Fig. 8) indicate that no striking effect on longevity is incurred even at 20% $D_2O$ enrichment during both embryonic and adult life (Strehler and Morgan, unpublished results).

The second group of findings deals with the rate of age pigment (lipofuscin) accumulation in Japanese as compared with Americans. It is known that this substance, many of whose properties have been described in collaboration with Hendley *et al.* (1963a, 1963b; Strehler *et al.*, 1959), accumulates at a constant average rate in the human myocardium (0.6%/decade), and that a similar pigment occurs in the uterine muscle of rodents reared on a diet deficient in the antioxidant, vitamin E (Strehler, 1964b). The observed accelerated rate of pigment accumulation in Japanese (Fig. 9) may result from a similar deficiency of vitamin E in oriental diets, or of a high level of unsaturated lipids in that diet. Both of these questions are susceptible to appropriate study.

Age pigments, as a prime probable example of crosslinkage by reactive intermediates produced in the course of noncatalyzed reactions (Strehler, 1963), would seem the ideal model for testing the crosslinkage hypothesis of aging. This may well be, but it must here again be reiterated that evidence is not yet available that these substances significantly impair function in the concentration at which they occur within long-lived mammals.[2] The effect of vitamin E level on pigment content and on the long-term survival of experimental mammals (preferably primates or dogs) is an important area for future systematic study.

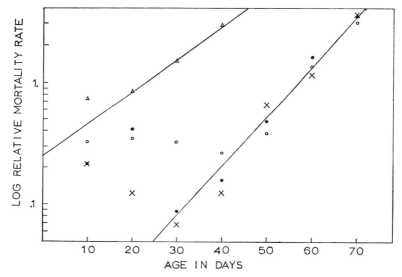

FIG. 8. Gompertz plot of mortality rate of Drosophila vs. age at different concentrations of $D_2O$ during the entire embryonic and adult life-span. Controls = ● ($N$ = 200); 10% $D_2O$ ($N$ = 200) = 0; 20% $D_2O$ = X ($n$ = 200); 33% $D_2O$ (n = 200) = ▲. After unpublished results of Strehler and Morgan.

A potentially important, but completely uninvestigated, possible source of age change exists in the related facts that (a) most biological subsystems subsist in an aqueous environment, and (b) that most, if not all, biosynthetic reactions, particularly those that generate high molecular weight components from simpler building blocks, take place through the net removal of a molecule of water at each addition to the growing polymer. Thus, ATP synthesis involves the net removal of water between ADP and $HPO_4^{2-}$; nucleic acids, proteins, carbohydrates, and fats are all synthesized through net dehydration reactions. Moreover,

---

[2] *Note added in proof:* Reichel, Hollaender, Clark, and Strehler have recently determined the fraction of intracellular volume occupied by lipofuschin in the neurons of the rat to be about 25% at 2 years of age. There would appear to be little doubt that this level of accumulation is harmful in its effect.

these synthetic polymers are not generated spontaneously, but, on the contrary, are joined to each other via routes of variable complexity in which the ultimate driving force is a reaction which yields a larger free energy than that consumed in the synthesis. It follows that most, if not all, such biopolymers are inherently unstable and that one of the most likely to occur of possible disruptive reactions in such systems is the hydrolysis of the peptide, 3′, 5′-phosphate, ester, or other linkage that binds the polymer together. To the author's knowledge, no attempt has been made to evaluate the extent of accumulation of partially hydrolyzed residues of such components during the aging of living systems, either in the presence or absence of specific endogenous hydrolases.

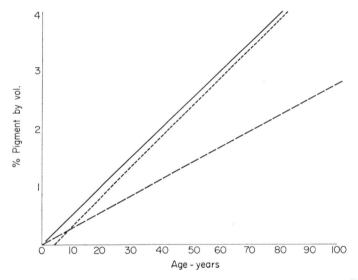

Fig. 9. Comparison of rate of accumulation of lipofuscin in Japanese (solid line) and American (dashed line) cardiac muscle. Average rate of pigment accumulation in Japanese autopsy specimens = 0.046% pigment volume/year ($n = 48$); in American sample = 0.028% pigment volume/year ($n = 156$).

Decreases in function, dependent on structural change, arise not only from disruptive reactions, both genetic and stochastic; they also occur whenever there is an imbalance between resynthesis (archaic: anabolism) and degradation (catabolism). Although genetically determined events may lie at the root of a substantial portion of the degradative changes that produce senescence, there are also ways in which an inadequacy of suitable raw materials for resynthesis can produce functional decline. Of particular importance may be nutritional deficiencies of various kinds.

Interestingly, one of the few experimental regimens that does produce a marked increase in the longevity of the laboratory rat is caloric restriction, as demonstrated by McKay et al. (1939) several decades ago. The mechanism of this effect, which consists of a 50–90% increase in longevity, is still not understood, but it is generally believed that a major portion of the increased life expectancy of McKay animals is due to a retarded rate of development (Barrows, 1966; Ross, 1959).

There is possibly a parallel effect in man. The life expectancy of humans is strikingly reduced by obesity (National Office of Vital Statistics, 1956; Keys, 1956). Much of this reduction in longevity may be ascribable to an increased incidence of coronary artery disease among the overweight, but other factors, including even some effects on the latter phases of development, may be implicated.

The foregoing observations imply that the ideal environment may not be the one in which food is overabundant or in which leisure is the rule. While no one doubts that malnutrition and undernutrition are harmful to life expectancy as well as to the expectancy of living, it may well be that superabundance is a worse enemy of longevity than moderate deprivation. This paradox may again be related to the doctrine set forth early in this paper: that a species is optimally adapted to the environment in which it evolved. If this is valid, it seems likely that our ancestors were a lean and hungry lot!

There are many cases in which enriched or supplemented diets increase longevity and in which specific deficiencies cause gross curtailment of life. Most of the evidence dealing with this important area, in which environmental manipulation may play an important role in determining both the length and quality of existence, has been treated elsewhere by workers more critically aware of the many faceted aspects of this problem (Barrows, 1966). There are, however, several instances of nutritional effects on life-span that should be reviewed briefly. These include the substantial increases observed by the late T. S. Gardner in the life-span of *Drosophila* (1948) when their diets are supplemented with royal jelly or certain of its separate components (pantothenate); the remarkable extension of longevity that the feeding of the same food produces in the potential queen bee; the increase in life-span of the domestic fly on a protein supplemented diet (Rockstein, 1959); and finally the substantial decrement in life-span which vitamin E deficiency engenders in the laboratory rat. As pointed out above, the last-named is potentially significant in its effect on lipofuscin accumulation. Systematic life tables for laboratory animals supplemented with various amounts of vitamin E have not yet been reported. However, deficiency of this factor is known to induce lipofuscin-like pigment secretion in the uteri of rats (Strehler, 1964b). Effects in other tissues have not been reported. Clearly, this is a fertile area for further studies. Life tables for laboratory animals under controlled conditions, though expensive and painstaking to obtain, are an essential part of the basic information relating environmental factors and the aging process.

### 3. Evolved Sources of Disruption: The Effect of Biological Factors

There are many ways in which organisms interact with each other both beneficially and detrimentally, and it is not always easy to predict what the effect on survival of a given interaction may be. Whatever the resolution of such biological encounters is at the sub- or supra-organismic level, it is clear that, given adequate time and a sufficiently predictable effect of the factor, there will be an appropriate evolutionary adaptation to it, sufficient to reduce its impact to proportions comparable to those posed by other challenges.

The Russian molecular biologist, Medvedev, has suggested (1964, 1966) that there may be a balance between the evolutionary advantage of substantial genetic variability (and consequently rapid evolutionary adaptability) and its disadvantages in terms of the stability of an individual organism. He has posed the thesis that there may exist mutator genes, genes controlling the synthesis, for example, of abnormal nucleotides, genes that program the failure of an organism as a by-product of a built-in and advantageous rate of error production. The occurrence of arabinosides (in sponges) and perhaps of other odd nucleotides may be related to this suggestion (Cohen, 1966).

The considerable body of data presented by Wulff and co-workers (1962, 1965, 1966) during the last decade on the age-dependent increment in nuclear RNA synthesis in a number of tissues of the rat has been interpreted by these workers as a possible reflection, as hypothesized by Medvedev, of the accumulation of such errors in DNA. Although such a thesis is only one of several plausible inferences from the data available, it is an interesting one and deserves further investigation.

Such autogenous errors due to aberrant raw materials exemplify molecular influences on functional capacity at a highly primitive level. More complex interactions arise from virus infections. The cataclysmic effects of lysogenic phage infections on the host bacterium are well documented (Lederberg, 1951), but it appears to be a rule that the most successful parasites are those which do not kill the host but rather make more minor demands on its resources. The highly successful intestinal flora are an example of this happy and frequent symbiosis. Studies on the life expectancy of completely germ-free animals (Gordon et al., 1966) have shown, contrary to Metchinhoff's auto-intoxication thesis (1907), that the common intestinal bacteria are not harmful to life expectancy.

Similar benign relationships have no doubt molded the evolution of species-specific viral parasites. Many of the viruses now believed to be important accompaniments, if not even exclusive causative agents in tumorogenesis (Hanafusa and Janafusa, 1966; Kelloff and Vogt, 1966; Temin, 1966), are by and large tolerated by the host as harmless intracellular cohabitants. These same autoreplicators, can when unleashed by as yet poorly understood events, produce the cataclysmic total destruction inherent in neoplasms. An analogy to the transformation of nonlysogenic into lysogenic bacteriophages is obvious.

If a significant fraction of cancerogenesis in humans can ultimately be traced to viral agents, then the type of molecular pathology an active virus generates is indeed an important contributor to the debilities of the aged; and suitable prophylaxis, immunization, or other public health measures may substantially reduce the unhappy consequences of this sadly ancient disease complex. The means employed by viral agents to commandeer the resources of the host cell, including particularly the protein and nucleic acid synthetic machinery, are major puzzles of no little import in the ontogeny of senescence. We have suggested (Strehler, 1966a) that viruses mobilize this synthetic apparatus by providing messages for proteins whose substrates are, on the one hand, sRNA species specifically suited to read other viral messages, and on the other hand, hydrolases that inactivate host-specific message translator agents (i.e., host sRNA's? See Section III,D).

A third class of molecular agent that may produce a portion of the senescence complex is that of toxic substances either endogenously produced or the vectors of interspecific competition. An important example of such interaction is the leaf-senescence inducing effect of certain plant hormones. Studies by Osborne (1962, 1966) have demonstrated that auxin applied in appropriate dosages has the capacity to retard senescence of the treated portion of *Euonymus* leaves while at the same time the treatment induced the accelerated senescence of leaf portions adjacent to the treated parts. Interestingly, the intermediate agent in this effect in certain species, but probably not in others, is the hydrocarbon, ethylene, whose production is also apparently moderated by auxin. This system of interactions is in many respects an ideal model for the study of programmed or induced senescence, and it is to be hoped that rapid further progress will be made in this important area of aging research.

Analogous in net result, but contrasting in mechanism, is the effect of the products of autoimmune reactions (Comfort, 1964). Autoimmune reactions are based on the failure to recognize self (Burnet, 1959; Walford, 1966). There are in principle at least two origins of such failure: the imprinting of self-tolerance on the immunological memory may gradually be lost or altered and allow the development of antibodies of normal body constituents; or alternately, the non-immunological system may liberate antigens to which tolerance was not established during the imprinting period, either because of the occurrence of mutations in cells which normally are exposed to the antibody-forming system or because of the breakdown of normal barriers between cell proteins and the host's immune defenses.

Variants of this hypothesis have enjoyed some vogue in recent years, although there is no evidence that autoimmune reactions occur universally in aged animals or humans. Indeed, various important diseases frequently regarded as manifestations of autoimmune reactions (e.g., lupus erythematosis, rheumatoid factor, circulating antineural antibodies, etc.) are frequently ascribable to prior damage to the organ involved, damage which exposes the host's defenses to a new array of its own constituent proteins, an array to which tolerance had not been developed during imprinting.

In summary, although there are many reports of the occurrence of autoimmune reactions at increasing frequency as a function of age, there is as yet no compelling evidence that this is a part of the normal course of aging; rather, the evidence seems to indicate that stochastic, genetic, or environmental events are its progenitors.

The interaction of cells within an organism through both hormonal and nutritional channels is of prime importance in maintaining function. A considerable body of evidence has become available regarding the age-dependent changes in levels of hormones, substances definable in terms of the present context, as chemical compounds synthesized by one cell of an organism, but producing a response (either stimulatory or inhibitory) in another. It is known that in some cases (e.g., androsterone) the production of hormone decreases with age; in other cases (pituitary gonadotropin) the level may actually increase with age until very late in life (Albert et al., 1956; Pincus, 1956).

There is some evidence bearing on the question of nutritional interdependence among different body tissues (Havel, 1965a; Pozefsky et al., 1965). Thus, liver, muscle, brain, adipose tissue, etc., are coordinated in a remarkably coherent series of syntheses, whose elements are just now being clarified. With certain notable exceptions, little is known of age-dependent alterations in such cross-dependence as a function of age. Striking decreases in glucose tolerance have been reported in humans (Havel, 1965b); decreased production of fatty acids in response to epinephrine administration has been noted in rats (Hruza and Jelinkova, 1965); and in the same species a diminished mobilization of dermal collagen reserves into liver glycogen stores in response to cortisol treatment has been observed (Hauck et al., 1966). These representative studies make it abundantly clear that cellular interdependences are modified during senescence, but the mode of this altered interaction is not yet worked out and may indeed be substantially different in each case.

Evidence presented 40-50 years ago by Carrel (1912) and Carrel and Ebeling (1923) regarding the effects of sera of different ages on the growth of cells *in vitro* was suggestive of a generally altered milieu. These experiments have not been adequately confirmed in detail; the nature of the substances responsible for growth stimulation or inhibition have not been clarified, and the suggestive effects of repeated plasmaphoresis on the aging of dogs has not been verified since Carrel's early reports.

The interaction of cells in such a manner as to produce the senescence and death of an organism is in a sense a surprising event for which no facile evolutionary rationalization can be advanced. If, as appears to be the case in certain instances, there is a program of senescence and death [e.g., in *Campanularia flexuosa* (Strehler, 1961) salmon (Beverton and Holt, 1959), epidermis (Montagna, 1962), erythrocytes (Brock, 1960)], such a program must arise from one of the following: (a) insufficient selection pressure to remove the source of deterioration in the ancestral population—perhaps because it manifests itself after adequate reproduction has taken place; (b) direct selection of the senescent program in order to better adapt the species to a variable niche (e.g., *Campanularia?*) or to provide for the nutrition of a new generation (salmon?); or (c) indirect selection (Birren, 1960; Medawar, 1951; Williams, 1957) of an evolutionarily slightly disadvantageous senescence as a by-product of a process or property selected because of its direct and considerable advantage. Fundamental information on the biology of cellular death is needed in order to appraise the role of this process in senescence models. A thorough biochemical profile and comparison of changes in a variety of cells that die naturally (e.g., epidermis) as a part of their ontogenetic program or in response to environmental assault (ultraviolet or X-radiation, burns, freezing, toxic agents, anaerobiosis, starvation, etc.) is very much needed, both in histochemical (Raychaudhuri et al., 1965; Strehler et al., 1966, 1967) and biochemical terms (Batra and Strehler, 1967; Booth et al., 1964).

That environmental factors directly attributable to other living systems are crucially important in determining the length and quality of life is readily apparent from a comparison of the mortality statistics of developed and underdeveloped cultures and countries (United Nations Demographic Yearbook, 1956). A paradoxical feature of the survivorship curves thus derived is that the less hospitable natural environments are not necessarily those that have produced the lower standards of living or higher age-specific mortality rates. A hostile environment, particularly one in which there is an absolute requirement for warm clothing, shelter, and appreciable food storage to last through the winters, may yield a higher standard of living than a tranquil, unchallenging tropical island where food is plentiful but where population can expand unprudently and eventually yield an overcrowded, overpopulated, improvident, listless paradise. Such difficulties can also emerge in barren wasteland, but in such environments the rigors of climate and nature can more readily eliminate those who are foolish nonvirgins or prodigal sons.

The interorganismic factors that influence survivorship curves consist, in addition to parasitism and predation, of various kinds of competition. The effect of competition on the mortality behavior of an experimental animal, *Drosophila*, is shown in the comparison of "sterile" and "nonmicroorganism free" survival curves illustrated in Fig. 10 (Strehler, 1962b). That predation and disease are dominant factors in the mortality behavior of animals in the wild is illustrated by the comparison of mean longevity and maximum longevity in birds. A robin may, under ideal conditions, return to the same locale for 12–20 years (Comfort, 1956). By contrast, the average life-span in the wild is less than 3 years.

Intraspecific competition among birds for space may be at least as rigorous as interspecific competition. This competition, and the forceful assertion of domain rights, limits the size of wild populations and no doubt accounts for some of the high mortality of young birds that have not successfully established domain rights.

The competition for space that appears so frequently among animals is probably a secondarily evolved manifestation of the competition for food and for sex. The individual who has effective and unchallenged control of a piece of nature suited to his food requirements and those of potential offspring is also in an effective position to mate and to compete effectively in the gene pool. The animal without a territorial fief is insecure economically, psychologically, and reproductively.

It is extremely difficult to sort out the environmental factors which affect the mortality behavior of a subpopulation, for both genetic and environmental influences are multiple and interdependent. Nevertheless, the insecurity of a propertyless, underfed, anchorless social group is probably a direct factor in the reduced mean longevity of such subpopulations. Psychological factors and related psychosomatic ills such as cardiac, hypertensive, and digestive disturbances affect longevity both directly and indirectly (*Medical World News*, 1965, 1966). Tense or alienated persons may compensate for their predicament by overindulgence in food, alcohol, or cigarettes, not to mention other unfortunate habits or personality attributes, unnecessary aggressiveness, or suicide. In a very real sense, the factors that determine length of life (with the possible exception of a superabundance of food) are those that lead to a fulfilled life. Few nongenarians are as irascible as G. B. Shaw, and a biologically long life probably is mainly engendered by habits of mood, sleep, and feeding that are instilled early in life and nurtured by that sense of security in a prescribed domain that home and family mean.

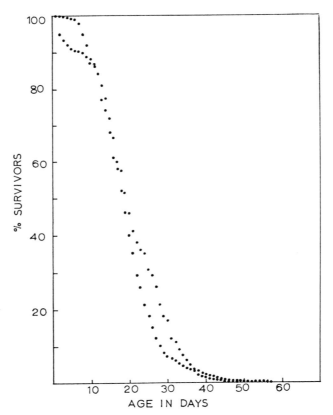

FIG. 10. Effect of aseptic conditions on survivorship curves of *D. melanogaster*. (●) Sterile conditions ($N = 1020$); (○) controls ($N = 3740$) (25°C). After Strehler (1962b).

The history of the human race and of other aggressive species is to a considerable extent the account of the drive man has to control his immediate milieu—to acquire and keep a domain. It seems clear that much of the turmoil that modern history has recorded is an expression of those primary needs for space, for food, and for sex that motivate so many nonhuman species. These urges may be primitive, but it is dubious that they are either unnecessary for long-term survival or capable of being ignored.

The instinctive response of the parent to the neighborhood bully, the cheers for the hometown football or baseball team, the pride in "my" country, or the pleasant excitement of seeing a man, American or Russian, black, yellow, or white, exploring outer space or the moon's craters, are simply imaginary extensions of the individual's domain, and the security generated as its borders expand.

Although the competition for space, food, and sex, with its automatic implications for survival and longevity, is a reality that will hopefully remain with future man, the minimization of its unnecessarily harmful effects—wars, aggressions, military expenditure, and the opposing philosophies which are generated by differential status in the control of available land, food, and sex (and particularly by the unresolved issues of who will control those areas with conflicting loyalties or histories)—is possible in even an imperfect world. To achieve this millenium, a substantial revision of old cliches about self-determination, majority rule, historical precedent, and the inheritance of land use may be required.

In the meantime, even simple written or philosophical agreement regarding the principles that will guide future decisions of human domain rights is lacking!

Wars and their long-term effects have very critical influences on life expectancy. Wars as modifiers of life tables or accelerators of senescence are less attractive as tools of experimental gerontology than most other agents discussed above. That the life of the soldier is not conducive to longevity is shown by the comparison of mortality rates (of soldiers and of civilians) from a considerable variety of causes as compiled by Jones (1956). In a real sense, an effective statesman may contribute more in practical geriatric effects than a small army of talented scientists and gentle physicians.

In summary, the complex and interdependent evolved factors, biological, psychological, and social, including man himself and drives for land control and the food and sex it provides, that operate in inter- and intraspecific competition have profound but incompletely understood effects on the longevity of man in the wild or domesticated state.

## III. ON THE POSSIBLE MODIFICATION OF THE RATE OF AGING THROUGH MANIPULATION OF THE INTERNAL ENVIRONMENT

The last part of this paper deals with several approaches which suggest themselves as potential means of changing the rate of aging. All of these potentialities are based on modification of factors that control the program of development, particularly the late-active parts of that program; all are based on the assumption that such a program does in fact exist and can be modified. They differ in the means proposed to modify program and in their speculative extremity. Specifically, they deal with the following possibilities: (a) that the onset or rate of senescence of higher mammals may be retarded by controlled dietary restriction (in analogy with the McKay results on rodents); (b) that a similar effect may be achieved through the deliberate reduction of the body temperature of warm-blooded animals (man); (c) that the genetic program may be retarded through the application of radiation at certain crucial phases of development; and (d) that it may be possible to slow down the schedule of aging (or even to reverse certain phases of it) through pharmacological intervention in the program readout machinery. We shall consider those possibilities in some detail.

### A. Controlled Growth and Maturation via Dietary Restriction

The McKay experiment has generally been interpretated as a prolongation of the developmental or growth period. In the absence of sufficient amounts of energy-yielding foods, animals grow at a substantially lower rate and continue this slow increment long after the control animals have ceased to grow. What missing dietary components are responsible for this effect is not known with certainty. Studies by Ross (1959) indicate that either carbohydrate or protein restriction will produce the effect. Similar results have been reported for rotifers (Fanestil and Barrows, 1965) and for the guppy (Comfort, 1961), but unfortunately, since the original McKay studies, no controlled measurements have been carried out on species closer to man than the laboratory rat. The recent increase in emphasis on primate research and the establishment of well-appointed and staffed centers for such research would seem to furnish opportunity to determine whether the McKay effect is likely to be observed in man. Because the effect in rodents is so striking (essentially a doubling of the mean life expectancy) even a partial parallelism in humans might add many useful years to life (provided that such a regimen didn't generate other problems of a more serious nature than a life-span limited to a mere 70 or 80 years). An "average" life-span of 70 years is probably preferable to a "mean" life-span of 85!

### B. Altered Body Temperature

As discussed earlier, the 1917 Loeb and Northrop studies on *Drosophila* established that the rate of senescence in this species is temperature dependent. Sufficient parallelisms in studies of other species of insects, of rotifers (Fanestil and Barrows, 1965), and of fish in the wild (Gerking, 1960), suggest that this is a general phenomenon that even expresses itself in vertebrates. However, once again, in the vertebrate of greatest interest to man (man), no evidence is available, for homeothermic mammals and birds rigorously maintain their body temperatures within a 1°–1.5°C range of values. Little is known of the capacity of the mammal to adapt to temperatures lower than "normal" for extended periods of time. Although more than average torpor is reportedly characteristic of dogs that have been kept at somewhat reduced internal temperatures, it cannot be concluded that such would be the case of primates or humans suitably adapted to a body temperature of 34° or 35°C for a longer period of time. Whether such adaptation is feasible, what its effects might be on other functions, particularly mental function and finally, what its effect might be on life expectancy are as yet unanswered and intriguing questions. The disadvantages of a poikilothermic existence as compared with a homeothermic one are not readily apparent, since chelonians (Comfort, 1956) have very extended life-spans and the oceans have not yet been taken over by mammals. The speed of movement of reptiles or amphibians is not less than that of mammals existing at 10°C higher temperatures, nor do rodents or birds appear to be more philosophically inclined than the reptiles that prey on them and digest them in contented, leisurely fashion.

It is therefore suggested that the advantage of homeothermy is primarily that maintenance of constant body temperature in the range of 38°C increases the potential for sustained work that can be incorporated into a given biological mass. The thermostat of homeotherms is apparently set near the upper range of thermal stability of their constituent proteins. At this temperature the turnover number of enzyme-catalyzed reactions is maximized. At somewhat higher temperatures rapid damage will result with consequent shortening of life or the obligatory development of a complex hierarchy of systematic cell replacements. At somewhat lower temperatures the capacity to do sustained work, particularly muscular work connected with food catching, predator avoidance, and combat, would be materially decreased to the deteriment of the low performers.

Our thesis, then, is that a mammal such as man, in a modern technological milieu, is no longer dependent on his capacity for sustained muscular work, unless he is a football player, mountain climber, or exceptionally affectionate; and that one of the most promising ways of adding years to life without subtracting life from years may be to find a means to reset the hypothalamic thermostat without producing idiots or sleepwalkers.

There are a number of drugs available for experimental or therapeutic use, capable of altering the thermostat in experimental animals. Twelve such compounds "with no major side effects" have been offered to the writer by one large drug house for use in experimental studies on longevity of the laboratory rat. Space and financial limitations have prevented our undertaking these studies thus far.

The problem may also be attackable in humans in a demographic format. No studies of the relation of human longevity pedigrees and thermostat setting pedigrees are available; nor do we know the effects of lingering illnesses that cause prolonged elevations of temperature (e.g., malaria) on the relative age-specific mortality rates of persons who have been treated and have recovered after many years of exposure; no careful studies of the subsequent mortality effects of special occupations that expose the worker to prolonged periods of hyperthermia are yet available; and even the effects of periodic hibernation on the longevity of hamsters or dormice have not been described, although Hruza et al. (1966) have demonstrated that the rate of collagen aging (crosslinkage) is reduced during hibernation of the latter. In short, this potentially rich area for investigation of a controllable environmental variable is not being effectively exploited.

## C. Interference with Genetic Program-Radiation

Ionizing radiation, as discussed in an earlier part of this paper, is primarily a disorganizing agent. It produces damage to a variety of structures including particularly the genetic apparatus. It is therefore surprising that X-rays are capable of extending the life span of several species, particularly of *Drosophila*

(Strehler, 1964a) and of the marine hydroid, *Campanularia flexuosa* (Strehler, 1964a) (see Fig. 11). Even in mammals, as indicated both by the work of Gowan (1960) and Sacher and Trucco (1962), the exposure of animals to small doses of X-rays may increase the life-span (Fig. 12). In Gowan's experiments, the explanation advanced was that the stresses of litter bearing were reduced by X-radiation (since the animals were sterilized or their litter sizes reduced). However, Sacher obtained similar results on male animals and ascribed the effect to a decrease in physiological variability or to nonspecific benefits. In our studies of

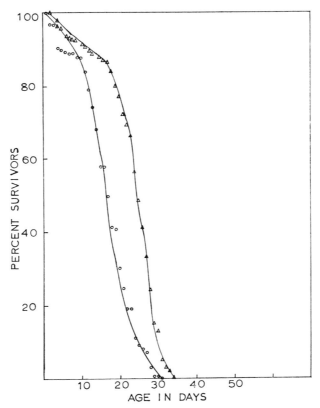

FIG. 11. Effect of ionizing radiation on survivorship curves of (a) *D. melanogaster* (radiation dosage about 5,000 r) and (b) *Campanularia flexuosa* (Strehler, 1964a).

*Drosophila*, in which life expectancy was increased even at 5000 r, it is believed that suppression of competing or pathogenic bacteria was an important facet of the causes, but in *Campanularia*, where 50,000–100,000 r resulted in almost a doubling of longevity and where control and experimental animals were maintained in the same medium, no such explanation is likely. Rather, in this case (and perhaps in others mentioned) the effect is probably due to a reduction in the effectiveness of some genetic component that programs the senescence of the animal. The radiation of an animal (or treatment with a radiomimetic drug) during the time that such a hypothetical genetic factor is being activated should inhibit the syntheses that led to the harmful effect, particularly if the adverse effect is initiated by cell division, a process which is particularly sensitive to radiation.

Whether such death-programming genes exist in higher mammals has not been established. If such genes are factors in the limitation of life-span, it may be feasible to suppress their action by selective and timely radiation of those cells in which fatal genetic processes are initiated.

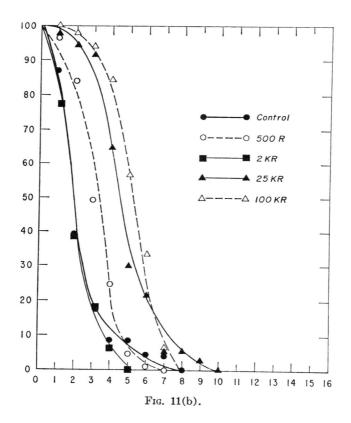

Fig. 11(b).

## D. Reversal or Suppression of Genetic Program: Pharmacologic Agents

During the past year we have presented a molecular-genetic model of senescence (Strehler, 1966a) in which it is postulated that aging is a by-product of limitations in syntheses that accompany the process of cell specialization (differentiation). This theory, derivative of the Weismannian (1891) and Minotian (1907) theory that aging is a consequence of, accompaniment of, or "price paid" for cellular differentiation, in its present formulation is based on the following specific theses:

(1) That messages coding for proteins characteristic of a given cell type employ

a specific and exclusive set or combination of codons, which set is not employed by any other cell type.

(2) That the kinds of messages a cell can translate into proteins are determined by the kinds of codon-specific sRNA-amino acid ligands that cell is capable of synthesizing.

(3) That differentiation occurs through the selective production of such codon-specific sRNA-amino acid ligands as are required to translate messages characteristic of the cell specific proteins.

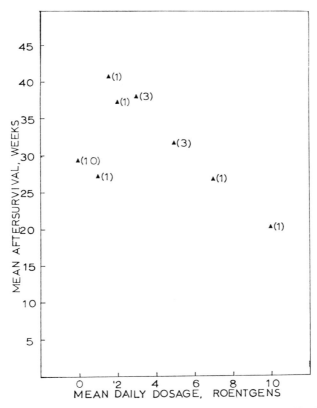

FIG. 12. Evidence that small doses of ionizing radiation lengthen life. The mean aftersurvival time of mice given small dosages of ionizing radiation is greater than the controls at doses between 2 and 4 r/day during the entire lifetime. (Parentheses indicate numbers of independent replication groups.) Sacher and Trucco (1962).

(4) That the gradual imposition of selective syntheses implicit in this model of differentiation eliminates earlier syntheses. If some of these newly untranslatable messages code for products that have very long half-lives but which, unless replenished, eventually deteriorate, a gradual result will be decreased function, cell death, and senescence.

This concept can economically be illustrated as follows. Suppose, for simplicity, that all proteins were made of only these amino acids, alanine, leucine, and serine, and that there are two code words that can be used to symbolize each amino acid, A, a, L, l, S, and s. It would now be possible to specify exclusively the proteins of eight ($2^3$) different cell types thus: ALS, ALs, Als, AlS, als, alS, aLS, and aLs.

If each of the eight cell types can read *only* the codons designated in its unique language, it will be able to read completely *only* messages written in that language. In other words, the translating ability of a cell determines the proteins it can synthesize from among all the protein specifications stored in its DNA.

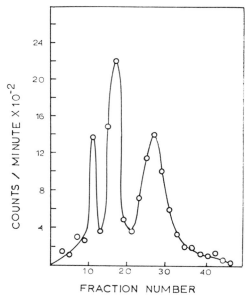

Fig. 13. Distribution of leucyl-sRNA synthetase activity in fractions of heart supernatant fluid eluted from DEAE cellulose. Three ml of heart 105,000g supernatant fluid was passed through a Sephadex G-25 column, applied to a 1.0 × 13.5 cm DEAE-Cl⁻ colunm, and eluted with 225 ml of a nonlinear NaCl gradient. ○ = 0.45 M NaCl, 0.05 M in Tris-Cl⁻ pH 7.5, 0.01 M mercaptoethanol. Assay: 50-μl aliquots of enzyme fractions were incubated with 20 μg rabbit liver sRNA and (final concentrations of) 0.01 mM l-leucine-³H, 5 curies/mmole, 0.005 M ATP, 0.005 M MgCl₂, 0.05 M potassium cacodylate buffer pH 7.0, 0.003 M GSH, and 0.0005 M Na₂EDTA. Total volume = 75 μl. After 30 minutes at 28°C, a 50-μl aliquot was applied to a filter paper disc, precipitated, washed, and counted. After Strehler (1966b).

If during differentiation of a given type of cell, the ability is lost to read a certain word (e.g., "a"), all syntheses employing messages specifying "a" will cease. Further, if some message specifying "a" produces a necessary product (x) with a limited half-life, those cells unable to translate the code word, "a," will fail as the product, "x," disappears. That is, the senescence of that cell type was determined when the ability to read "a" was lost.

During the past 18 months, Dr. D. Hendley, Mr. G. Hirsch, and the author have been attempting to test aspects of the first three theses. Although specific sRNA-AA complex formation could theoretically be limited by the absence of either the substrates (sRNA, AA) or of the enzymes (sRNA ligases) involved, we decided initially to explore the latter possibility. We have obtained the following evidence, germane to the theory, thus far:

(a) Mammalian tissues (rabbit heart) contain several chromatographically separable leucyl-sRNA ligase activities (Strehler, 1966b) (see Fig. 13).

(b) These separable enzymes attach leucine to chromatographically separable species of sRNA (Strehler et al., 1966b) (see Fig. 14).

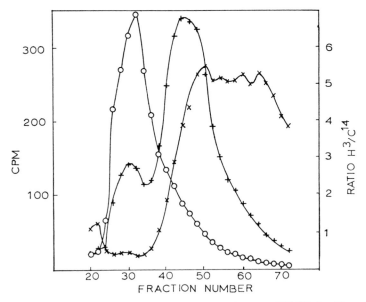

Fig. 14. Elution patterns of leucyl-sRNA's charged by two DEAE fractions from heart, then mixed, and co-chromatographed on an MAK column. (+) = $^3$H-leucyl-sRNA, formed by Peak I enzyme. (○) = $^{14}$C-leucyl-sRNA, formed by Peak III enzyme, (×) = ratio, $^3$H/$^{14}$C. Peak I enzyme represents pooled fractions 10–12 and Peak III represents fractions 24–31 (Fig. 13). The pooled fractions were concentrated by precipitation with 80% saturated $(NH_4)_2SO_4$, pH 7.2, 0.01 M in mercaptoethanol. After Strehler (1966b).

(c) Certain tissues (rabbit liver, reticulocytes) contain ligase activities specific to certain sRNA fractions with alanine acceptor activity; the same ligase activities are absent in other tissues (rabbit kidney) (Strehler et al., 1967) (see Fig. 15).

These results, although certainly not conclusive proof of the suggested model of differentiation, were predicted by the model and increase the likelihood of its relevance to developmental processes. Although further work will be required to establish its relevance to senescence or lack thereof, were aging to be programmed by systematic restrictions during development in the kinds of messages that can be read such as are here postulated, it should not be impossibly difficult to introduce pharmacologic agents that could either specifically retard or even reverse the effects of programmed codon usage restrictions or activations. Among the potential mechanisms for codon usage repression would be the production of enzymes that specifically destroy sRNA corresponding to a given codon useful at an earlier stage of development. Such a lytic activity might well be inhibited by (a) suitable analogues to the moiety normally affixed to the combining site on the enzyme; (b) the introduction of pharmacologically active substances that allosterically broaden the specificity of ligases present in the cell to include sRNA anticodons that are forbidden in the differentiated state; and (c) infection

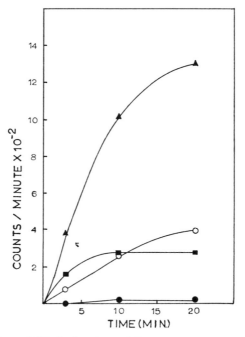

FIG. 15. Incorporation of $^3$H-alanine into sRNA fractions 125 and 150 from a Kelmer column, by crude enzyme preparations from reticulocytes and kidney. (○) = fraction 125 sRNA, kidney enzyme; (▲) = fraction 125 sRNA, liver enzyme; (●) = fraction 150 sRNA, kidney enzyme; (■) = fraction 150 sRNA, liver enzyme. After Strehler (1966b).

with specifically modeled "viruses" that code for the syntheses of missing ligases, etc.

The complex and unpredictable behavior of new pharmacological preparations, not to mention ancient remedies still in use, suggests that an "elixir" may not be as fantastic an alchemist's dream as were the dreams of elementary transformation in the years before Szilard, Fermi, and Los Alamos or the dreams of space travel in the decades before the Russian sputniks. If the programming of senescence is relatively simple at the genetic level, then no doubt its undoing through environmental control will also not be too improbable.

## APPENDIX I
### Evaluative Summary of Evidence on the Basic Hypotheses of Biological Aging

| Hypothesis No. | A<br>Statement of hypothesis | B<br>Model experiments or questions | C<br>Relevant findings | D<br>Tentative conclusions | E<br>Literature references |
|---|---|---|---|---|---|
| | *Senescence is due to* | | | | |
| | Class I: Loss of cellular syntheses due to: | | | | |
| | A. Genetic damage | | | | |
| (1) | 1. DNA mutation (or breakage) | Does increase in accumulated mutations decrease lifespan proportionally? | No! | Less than 10% of aging is due to this | Curtis, 1966; Clark and Rubin, 1961; Strehler, 1964a |
| (2) | 2. Crosslinkage | Does crosslinkage of DNA occur, or Do DNA crosslinkers accelerate aging? | To slight extent<br>No evidence | Need experiments<br>Need experiments | von Hahn, 1964, 1965 |
| | B. Loss of message synthesis | | | | |
| (3) | 1. Inhibitor-repressor accumulation (e.g., histones) | Are there substances in old cells that repress message synthesis by young DNA (+ enzymes)? | Histones are bound to DNA; metal ions are bound to DNA | Extent of young/old difference in histone binding makes mechanism unlikely | von Hahn, 1964, 1965 |
| (4) | 2. Absence of adjunct factors (e.g., sRNA, sRNA polymerase | Are ribosomes, sRNA's activating enzymes of old tissues capable of sustaining synthesis with young DNA as template? | None | Need experiments | |

| | | | | | |
|---|---|---|---|---|---|
| | C. Loss of message translating ability | | | | |
| (5) | 1. Unspecific loss of protein synthetic capacity | Same as (4) | None | Need experiments | |
| (6) | 2. Loss of ability to translate certain messages (codon restriction hypothesis) | Are there tissue differences in specific sRNA acylases? Do these differences result in tissue specific synthetic losses? | Yes | May be of considerable importance. Need experiments | Strehler, 1966a; Strehler et al., 1967a |
| (3 or 6a) | a. Specialized products | Do cells of different types differ in codon reading specificities? | Apparently | This model (or No. 3) may account for large percentage (60–95%) of decline | Strehler, 1966a; Strehler et al., 1967a |
| (3 or 6b) | b. Unspecialized cell components (e.g., membranes) | | | | |
| (3 or 6c) | c. Mitotic components | Do specialized syntheses repress type b and c syntheses? | Usually | Active exploration needed | Hay and Strehler, in press; Hayflick, 1965; Puck et al., 1958; Strehler et al., 1963 |
| | Class II: Loss of cellular function due to: A. Deterioration of semiautonomous or symbiotic organelles | | | | |
| (7) | 1. Mitochondria | Are mitochondria of old tissues identical with young? | No, they are more readily uncoupled. However, P/O ratios of fresh mitochondria unaltered during aging (rat) | Unlikely source of senescence | Barrows et al., 1960; Gold et al., 1966; Sallman et al., 1966; Weinbach, 1959 |
| (7a) | 2. Plastids | Are plastids of old tissues identical with young? | | Evidence not available | |

APPENDIX I (*Continued*)

| Hypothesis No. | A Statement of hypothesis | B Model experiments or questions | C Relevant findings | D Tentative conclusions | E Literature references |
|---|---|---|---|---|---|
| (8) | B. Insufficiency of promoter substances, e.g., hormones, growth factors, etc. | Do systemic promoting factors limit old organisms? Does parabiosis, transplantation or transfusion alter cellular age effects? | Yes, in some cases, e.g., growth hormone; factors in serum or embryos, 11-keto, 17-deoxy steroids; pituitary changes evident in rats. Diabetes increases vs. age | Probably important. May account for 20–80% of decrement in mammals | Osborne, 1962, 1966; Carrel, 1912, 1924; Kutsky and Feichtmeir, 1962; Pozefsky et al., 1965 |
|  | C. Accumulation of inhibitors |  |  |  |  |
|  | 1. Systemic (soluble components) |  |  |  |  |
| (9) | a. Serum factors (lipid peroxides) | Do inhibitory substances accumulate? Does transfusion of old to young shorten life or vice versa? | Probably, according to Carrel. Probably: plasmaphoresis | Suggestive: may contribute substantially | Carrel and Ebeling, 1923; Barber and Wilber, 1959 |
| (10) | b. Autoantibodies | Do old animals have increased autoimmune antibodies? Is this universal? | Yes<br><br>No | Specific lesions<br><br>Not a general process | Walford, 1966 |
|  | 2. Extra- or intracellular residues |  |  |  |  |
| (11) | a. Symbiotic viruses | Do symbiotic viruses decrease function of mammals? | Not known: many tissues, however, are virus-laden at advanced ages | Need experiments | Smith, 1966; Kelloff and Vogt, 1966 |

| | | | | |
|---|---|---|---|---|
| (12) | b. Exogenous precipitates (e.g., Ca salts) | Do such precipitates occur? | Yes, in certain diseases but probably not in normal aging | Unlikely to be important basic cause | Solomon, 1966; Lansing *et al*, 1950 |
| (13) | c. Crosslinkages | Do crosslinked materials occur? | Yes, age pigments; collagen is crosslinked, old collagen is not readily mobilized by aged rats (cortisol) | May be important impairment. Reduces energy reserve available | Verzar and Thoene, 1960; Hendley *et al*, 1963b; Strehler, 1964b; Hauck *et al*, 1966; Chvapil and Hruza, 1959 |
| | d. Thermal denaturation (catalyzed) | Are there accumulations of denatured material? | No evidence | Need experiments | |
| (14) | | Does heat shock decrease life expectancy of survivors? | No, in *Drosophila* | Probably not important | Morgan and Strehler, unpublished; Loeb and Northrop, 1917; Clarke and Smith, 1961 |
| | | Are enzymes damaged *in vivo* by reactions they catalyze? | Probably in some cases, but extent *in vivo* unknown | Need evidence | |
| (15) | e. Hydrolyses | Are partially hydrolyzed, long-lived residues present? | No evidence | Need experiments | |
| (16) | f. Isomerization | Does old protein contain excess D-amino acids? | Probably not | Unlikely source; more testing needed | Kuhn, 1958 |
| | Class III: Loss of intercellular coordination due to: | | | | |
| | A. Physical alterations in | | | | |
| (17) | 1. Membrane permeability | Does membrane permeability decrease with age? | In some cases permeability increases. | Unlikely source of aging | Kirk and Laursen, 1955 |
| | | Will agents that reduce permeability cause life shortening? | Radiation causes arterior-capillary fibrosis, shortens life | May be highly important Both above need critical tests | Casarett, 1964 |

APPENDIX I (Continued)

| Hypothesis No. | A<br>Statement of hypothesis | B<br>Model experiments or questions | C<br>Relevant findings | D<br>Tentative conclusions | E<br>Literature references |
|---|---|---|---|---|---|
| (18) | 2. Spatial relations between cells | Does cell slippage occur? Does increased cell movement or dislocation shorten life? | Evidently. Old tissues frequently recognizable by irregularity of spacing | Probably of some import | Strehler, 1964 |
| (19) | 3. Viscosity of diffusion medium or diffusion path | Is diffusion rate altered? | Not clearly known | Need experiments on tissue diffusion rates vs. age and viscosity measurements | |
| | B. Decreased cell responsiveness | | | | |
| (20) | 1. Loss of receptor sites on cells | Do old cells show decreased responses? | Yes | Probably important | Albert et al., 1956; Pozefsky et al., 1965; Krohn, 1962 |
| (21) | 2. Loss of transducing machinery | Is specific enzyme machinery decreased? | Not known | Need experiments | |
| (22) | 3. Loss of energy reserves | Are reserves decreased? | Free fatty acid release via epinephrine decreased | Important sources of decreases in sustained response | Hruza and Jelinkova, 1965 |
| | | | Collagen mobilization via cortisol decreased vs. age | Information on glycogen mobilization and re-synthesis needed | Hauck et al, 1966 |
| | C. Cell loss | | | | |
| (23) | 1. Accidental | Do cell losses occur? | Yes, 10–40% in some tissues | Probably not main source of dysfunction of muscle, heart, skin, liver, kidney; may be significant in nervous tissue and endocrines. Studies on mechanisms of cell death needed | Strehler et al., 1966, 1967; Raychaudhuri et al., 1965; Ellis, 1920; Birren and Wall, 1956; Brody, 1955 |
| | 2. Programmed | Does increased cell loss increase mortality? | Not known without complicating factors | | |

## REFERENCES

Albert, A., Randall, R. V., Smith, R. A., and Johnson, C. E. (1956). Urinary excretion of gonadotropin as a function of age. In "Hormones and the Aging Process" (E. T. Engle and G. Pincus, eds.), pp. 49–57. Academic Press, New York.

Alexander, P. (1966). The role of DNA lesions in the processes leading to ageing in mice. Proc. S.E.B. Symp. Aging, Sheffield, England 1966 (H. W. Woolhouse, ed.). Cambridge Univ. Press, Cambridge, England.

Alpatov, W. W., and Pearl, R. (1929). Experimental studies on the duration of life. XII. Influence of temperature during the larval period and adult life on the duration of life of the imago of Drosophila melanogaster. Amer. Naturalist 63, 37–67.

Atwood, K. C., and Pepper, F. J. (1961). Erythrocyte automosaicism in some persons of known genotype. Science 134, 2100–2102.

Barber, A. A., and Wilber, K. M. (1959). The effect of x-irradiation on the antioxidant activity of mammalian tissues. Radiat. Res. 10, 167–175.

Barrows, C. H. (1966). Enzymes in the study of biological aging. In "Perspectives in Experimental Gerontology" (N. W. Shock, ed.), pp. 169–181. Thomas, Springfield, Illinois.

Barrows, C. H., Jr., Falzone, J. A., and Shock, N. W. (1960). Age differences in the succinoxidase activity of homogenates and mitochondria from the livers and kidneys of rats. J. Gerontol. 15, 130–133.

Batra, P., and Strehler, B. L. (1967). Studies on the mechanism of cellular death. IV. Affect of anaerobiosis on ATP level and oxidative phosphorylation rate in mouse heart in vitro. Gerontologia 13, 30–36.

Beverton, R. J. H., and Holt, S. J. (1959). Review of the lifespans and mortality rates of fish in nature. In "The Lifespan of Animals," CIBA Foundation Colloquia on Aging, Vol. V. Little, Brown, Boston, Massachusetts.

Birren, J. E. (1960). Behavioral theories of aging. In "Aging: Some Social and Biological Aspects" (N. W. Shock, ed.), pp. 305–332. Publ. No. 65, American Association for the Advancement of Science, Washington, D. C.

Birren, J. E., and Wall, P. D. (1956). Age changes in conductive velocity, refractory period, number of fibers, connective tissue space, and blood vessels in sciatic nerves of rats. J. Comp. Neurol. 104, 1–16.

Bjorksten, J. (1958). A common molecular basis for the aging syndrome. J. Amer. Geriat. Soc. 6, 740–748.

Blum, H. G. (1959). "Carcinogenesis By Ultraviolet Light." Princeton Univ. Press, Princeton, New Jersey.

Booth, B. A., Creasey, C., and Sartorelli, A. C. (1964). Alterations in cellular metabolism associated with cell death induced by uracil mustard and 6-thioguanine. Proc. Nat. Acad. Sci. USA 52, 1396–1402.

Brock, M. A. (1960). Production and life span of erythrocytes during hibernation in the golden hamster. Am. J. Physiol. 198, 1181–1186.

Brody, H. (1955). Organization of the cerebral cortex. III. A study of aging in the human cerebral cortex. J. Comp. Neurol. 102, 511–556.

Burnet, F. M. (1959). Clonal selection. Croonian Lecture, Royal College of Physicans of London.

Carrel, A. (1912). On the permanent life of tissues outside of the organism. J. Exp. Med. 15, 516–528.

Carrel, A. (1924). A diminution artificielle de la concentration des proteines du plasma pendant la vieillesse. C.R. Soc. Biol., (Paris) 90, 1005–1007.

Carrel, A., and Ebeling, A. H. (1923). Antagonistic growth principles of serum and their relation to old age. J. Exp. Med. 37, 653–658.

Casarett, G. W. (1964). Similarities and contrasts between radiation and time pathology. In "Advances in Gerontological Research" (B. L. Strehler, ed.), pp. 109–164. Academic Press, New York.

CHANCE, B., AND WILLIAMS, G. R. (1956). The respiratory chain and oxidative phosphorylation. *Advan. Enzym.* **17,** 65.
CHVAPIL, M., AND HRUSA, Z. (1959). The influence of aging and undernutrition on chemical contractility and relaxation of collagen fibres in rats. *Gerontologia* **3,** 241–252.
CLARK, A. M., AND RUBIN, M. A. (1961). The modification by x-rays of the life span of haploids and diploids of the wasp, *Habrobracon* species. *Radiat. Res.* **15,** 244–253.
CLARKE, J. M., AND SMITH, J. M. (1961). Independence of temperature on the rate of ageing in *Drosophila subobscura*. *Nature* **190,** 1027–1028.
COHEN, S. S. (1966). Introduction to the biochemistry of D-arabinosyl nucleosides. *In* "Progress in Nucleic Acid Research and Molecular Biology" (J. N. Davidson and W. E. Cohn, eds.), Vol. V, pp. 1–89. Academic Press, New York.
COMFORT, A. (1956). "The Biology of Senescence." Holt, New York.
COMFORT, A. (1961). The longevity and mortality of a fish in captivity. *Gerontologia* **5,** 209–222.
COMFORT, A. (1964). "Ageing: The Biology of Senescence." Holt, New York.
CURTIS, H. J. (1963). Biological mechanisms underlying the aging process. *Science* **141,** 68–94.
CURTIS, H. J. (1966). "Biological Mechanisms of Aging." Thomas, Springfield, Illinois.
CURTIS, H. J., AND GEBHARD, K. L. (1958). The relative biological effectiveness of fast neutrons and x-rays for life shortening in mice. *Radiat. Res.* **9,** 278–284.
"Dictionary of Physics," Vol. 2, p. 26. Pergamon Press, New York (1961).
ELLIS, R. S. (1920). Norms for some structural changes in the human cerebellum from birth to old age. *J. Comp. Neurol.* **32,** 1–34.
FAILLA, G. (1960). The aging process and somatic mutations. *In* "The Biology of Aging" (B. L. Strehler, ed.), pp. 170–175. Publ. No. 6, American Institute Biological Sciences, Washington, D.C.
FANESTIL, D. D., AND BARROWS, C. H., JR. (1965). Aging in the rotifer. *J. Gerontol.* **20,** 462–469.
FERRIS, B. G. (1966). Environmental hazards: Electromagnetic radiation. *New Eng. J. Med.* **275,** 1100–1105.
FRANCK, J., AND PLATZMAN, R. (1954). Physical principles underlying photochemical, radiation-chemical, and radiobiological reactions. *In* "Radiation Biology" (A. Hollaender, ed.), Vol. I, pp. 191–254. McGraw-Hill, New York.
GAFFRON, H. *et al.* (eds.). (1957). "Research in Photosynthesis." Wiley (Interscience), New York.
GARDNER, T. S. (1948). The use of *Drosophila melanogaster* as a screening agent for longevity factors. I. Pantothenic acid as a longevity factor in royal jelly. *J. Gerontol.* **3,** 1–8.
GERKING, S. D. (1960). Evidence of aging in fishes. *In* "The Biology of Aging" (B. L. Strehler, ed.). American Institute of Biological Sciences, Washington, D.C.
GOLD, P. H., NORDGREN, R., AND STREHLER, B. L. (1966). A re-examination of the efficiency of oxidative phosphorylation versus age in rat kidney, liver, and heart. *Proc. Int. Congr. Gerontol., 7th, Vienna, 1966.* Vol. 6, Wien Med. Akad., Vienna, Austria.
GORDON, H. A., BRUCKNER-KORDOSS, E., AND WOSTMANN, B. S. (1966). Aging in germ-free mice: Life tables and lesions observed at natural death. *J. Gerontol.* **21,** 380–387.
GOWEN, J. W. (1960). Lengthening of life span in mice in relation to two deleterious agents. *In* "The Biology of Aging" (B. L. Strehler, ed.), pp. 188–191. Publ. No. 6. American Institute of Biological Sciences, Washington, D.C.
HALL, D. A. *et al.* (1955). Collagen and elastin in connective tissue. *J. Gerontol.* **10,** 388–400.
HANAFUSA, H., AND JANAFUSA, T. (1966). Determining factor in the capacity of Rous Sarcoma virus to induce tumors in mammals. *Proc. Nat. Acad. Sci. USA* **55,** 532.
HARMAN, D. (1962). Free radical theory of aging: Prolongation of the normal life span by free radical inhibitors. *In* "Biological Aspects of Aging" (N. W. Shock, ed.), Columbia Univ. Press, New York.

HARMAN, D., AND PIETTE, L. H. (1966). Free radical theory of aging: Free radical reactions in serum. *J. Gerontol.* **21**, 560–566.
HARVEY, E. N. (1940). "Living Light." Princeton Univ. Press, Princeton, New Jersey.
HAUCK, J. C., DEHESSE, D., AND JACOB, R. (1966). Effect of aging upon collagen catabolism. *Proc. S.E.B. Symp. Aging, Sheffield, England, 1966* (H. W. Woolhouse, ed.). Cambridge Univ. Press, Cambridge, England.
HAVEL, R. J. (1965a). Autonomic nervous system and adipose tissue. *In* "Handbook of Physiology." Adipose Tissue 58. American Physical Society, New York.
HAVEL, R. J. (1965b). Metabolism of lipids in chylomicrons and very low density lipoproteins. *In* "Handbook of Physiology," Adipose Tissue 50, Sect. 5. American Physical Society, New York.
HAY, R., AND STREHLER, B. L. (1967). Limited growth span of chick embryonic fibroblasts *in vitro*. *Exp. Gerontol.* (in press).
HAYFLICK, L. (1965). The limited *in vitro* lifetime of human diploid cell strains. *Exptl. Cell Res.* **37**, 614–636.
HENDLEY, D. D., MILDVAN, A. S., REPORTER, M. C., AND STREHLER, B. L. (1963a). The properties of isolated human cardiac age pigment. II. Chemical and enzymatic properties. *J. Gerontol.* **18**, 250–259.
HENDLEY, D. D., MILDVAN, A. S., REPORTER, M. C., AND STREHLER, B. L. (1963b). The properties of isolated human cardiac age pigment. I. Preparation and physical properties. *J. Gerontol.* **18**, 144–150.
HOLLAENDER, A. (ed.) (1954). "Radiation Biology," Vol. I, Chapters 11, 14, and 15. McGraw-Hill, New York.
HRUZA, Z., AND JELINKOVA, M. (1965). Carbohydrate metabolism after epinephrine, glucose and stress in young and old rats. *Exp. Gerontol.* **1**, 139–147.
HRUZA, Z. et al. (1966). The effect of cooling on the speed of ageing of collagen *in vitro* and in hibernation of the fat dormouse. *Exp. Gerontol.* **2**, 29–35.
HUENNEKENS, F. M., AND WHITELEY, H. R. (1960). Phosphoric acid anhydrides and other energy-rich compounds. *In* "Comparative Biochemistry" (M. Florkin and H. S. Mason, eds.), Vol. I, pp. 107–181. Academic Press, New York.
JOHNSON, F. H. (ed.) (1955). "The Luminescence of Biological Systems." American Association for the Advancement of Science, Washington, D.C.
JONES, H. B. (1956). A special consideration of the aging process, disease, and life expectancy. *Advances Biol. Med. Phys.* **4**, 281–337.
KAPLAN, H. S. (1966). DNA-strand scission and loss of viability after X-irradiation of normal and sensitized bacterial cells. *Proc. Nat. Acad. Sci. USA* **55**, 1442–1446.
KELLOFF, G., AND VOGT, P. (1966). Localization of avian tumor virus group-specific antigen in cell and virus. *Virology* **29**, 377–384.
KEYS, A. (1956). Diet and epidemiology of coronary heart disease. *J. Chronic Dis.* **4**, 364–380.
KIRK, J. E., AND LAURSEN, T. J. S. (1955). Changes with age in diffusion coefficients of solutes for human tissue membranes. *In* "CIBA Foundation Colloquia on Aging," Vol. I ("General Aspects"), pp. 69–75. Little, Brown, Boston, Massachusetts.
KROHN, P. L. (1962). Heterochronic transplantations in the study of aging. *Proc. Roy. Soc.* [*Biol.*] **157**, 128–147.
KUHN, W. (1958). Possible relation between optical activity and aging. *Advances Enzymol.* **20**, 1–30.
KUTSKY, R. J., AND FEICHTMEIR, T. V. (1962). Mitosis-stimulating properties of embryonic nucleo-protein constituents in cell culture. *Nature* **194**, 1050–1051.
LANSING, A. I., ALEX, M., AND ROSENTHAL, T. B. (1950). Calcium and elastin in human arteriosclerosis. *J. Gerontol.* **5**, 112–119.
LEDERBERG, E. M. (1951). Lysogenicity in *E. coli* K-12. *Genetics* **36**, 560.
LOEB, J., AND NORTHROP, J. H. (1917). On the influence of food and temperature upon the duration of life. *J. Biol. Chem.* **32**, 103–121.

Lovelock, J. E. (1954). Biophysical aspects of the freezing and thawing of living cells. *Proc. Roy. Soc. Med.* **47,** 60.

Ma, C. K., and Cowdry, E. V. (1950). Aging of elastic tissue in the human skin. *J. Gerontol.* **5,** 203–210.

McElroy, W. D. (1947). Mechanism of inhibition of cellular activity by narcotics. *Quart. Rev. Biol.* **22,** 25–59.

McKay, D. M., Maynard, L. A., Sperling, G., and Barnes, L. L. (1939). Retarded growth, lifespan, ultimate body size, and age changes in the albino rat after feeding diets restricted in calories. *J. Nutr.* **18,** 1–13.

Maynard-Smith, J. M. (1959). A theory of ageing. *Nature* **184,** 956–958.

Medawar, P. B. (1951). "An Unsolved Problem of Biology," Lewis, London. (An inaugaral lecture delivered at University College, London.)

*Medical World News* (1965). The stressful life on "Animal Farm," pp. 91–92, September 10.

*Medical World News* (1966). Probing man's world to clarify his ailments, pp. 130–132, December 2.

Medvedev, Zh. A. (1964). The nucleic acids in development and aging. *In* "Advances in Gerontological Research" (B. L. Strehler, ed.), Vol. I. Academic Press, New York.

Medvedev, Zh. A. (1966). Molecular aspects of aging. *Proc. S.E.B. Symp. Aging, Sheffield, England, 1966.* (H. W. Woolhouse, ed.). Cambridge Univ. Press, Cambridge, England.

Meryman, H. T. (1956). Mechanics of freezing in living cells and tissues. *Science* **124,** 515.

Metchinkoff, E. (1907). "The Prolongation of Life-optimistic Studies." Heinemann, London.

Mildvan, A. S., and Strehler, B. L. (1960). A critique of theories of mortality. *In* "The Biology of Aging" (B. L. Strehler *et al.*, eds.), pp. 309–314. Publ. No. 6, American Institute of Biological Sciences, Washington, D.C.

Minot, C. S. (1907). *Popular Sci. Monthly* **71,** June, pp. 481–496; August, 97–120; December, 509–523.

Montagna, W. (1962). "Structure and Function of Skin," 2nd edition. Academic Press, New York.

Montagna, W. (ed.) (1965). *In* "Advances in Biology of Skin," Vol. VI ("Aging"). Pergamon Press, New York.

Moore, W. J. (1963). "Physical Chemistry," 3rd edition, p. 621. Prentice-Hall, Englewood Cliffs, N.J.

National Office of Vital Statistics (1956). Death rates by age, race, and sex, United States, 1950–1953; Selected causes. *Vital Statist.-Special Reports* **43,** Nos. 1–31.

Osborne, D. J. (1962). Plant physiology, *Lancaster* **37,** 595.

Osborne, D. J. (1966). *Proc. S.E.B. Symp. Aging, Sheffield, England, 1966* (H. W. Woolhouse, ed.). Cambridge Univ. Press, Cambridge, England.

Pincus, G. (1956). Aging and urinary steroid excretion. *In* "Hormones and the Aging Process" (E. T. Engle and G. Pincus, eds.), pp. 1–19. Academic Press, New York.

Pozefsky, T., Colker, J. G., Langs, H. M., and Andres, R. (1965). The cortisone-glucose tolerance test. *Annal Intern. Med.* **63,** 988–997.

Puck, T. T., Cierciura, S. J., and Robinson, A. (1958). Genetics of somatic mammalian cells. III. Long-term cultivation of euploid cells from human and animal subjects. *J. Exp. Med.* **108,** 945–956.

Rabinowitch, E. (1954). "Photosynthesis." Academic Press, New York.

Raychaudhuri, A., Strehler, B. L., Wilder, R. J., Gee, M., and Press, G. (1965). Studies on the mechanism of cellular death. II. Carbohydrate and lipid changes during early and late cardiac necrosis in the dog. *J. Gerontol.* **20,** 338–345.

Rockstein, M. (1959). The biology of ageing insects. *In* "The Lifespan of Animals," CIBA Foundation Colloquia on Aging, Vol. 5, pp. 247–264. Little, Brown, Boston, Massachusetts.

Ross, M. H. (1959). Protein, calories, and life expectancy. *Fed. Proc.* **18,** 1190–1207.

Rubner, N. (1908). Probleme des Wachstums und der Lebensdauer. *Mitt. Ges. Inn. Med., Wien* **7** (Suppl. 9), 58–81.

SACHER, G. (1956). On the statistical nature of mortality with special reference to chronic radiation. *Radiology* **67**, 250–257.

SACHER, G. A., AND TRUCCO, E. (1962). A theory of the improved performance and survival produced by small doses of radiations and other poisons. *In* "The Biological Aspects of Aging" (N. W. Shock, ed.), pp. 244–251, Columbia Univ. Press, New York.

SALLMAN, B., STARCK, R., AND DEVELASCO, F. A. (1966). Cardiac tissue metabolism in aging. *Proc. Intern. Congr. Gerontol., Vienna, Austria, 1966*. Vol. 6. Wien. Med. Acad., Vienna, Austria.

SINEX, F. M. (1957). Aging and the lability of irreplaceable molecules. *J. Gerontol.* **12**, 190–198.

SMITH, G. H. (1966). Role of the milk agent in disappearance of mammary cancer in C3H/StWi mice. *J. Nat. Cancer Inst.* **36**, 685–701.

SOLOMON, R. D. (1966). The biology and pathogenesis of vascular disease. *Advances Gerontol. Res.* **2**, 285–354.

STREHLER, B. L. (1952). Photosynthesis—Energetics and phosphate metabolism. *In* "Phosphorus Metabolism," Vol. II, pp. 491–502. Johns Hopkins Press, Baltimore, Maryland.

STREHLER, B. L. (1955). Factors and biochemistry of bacterial luminescence. *In* "The Luminescence of Biological Systems" (F. H. Johnson, ed.), pp. 209–255. American Association for the Advancement of Science, Washington, D.C.

STREHLER, B. L. (1959). Origin and comparison of the effects of time and high-energy radiations on living systems. *Quart. Rev. Biol.* **34**, 117–142.

STREHLER, B. L. (1961). Studies on the comparative physiology of aging. II. On the mechanism of temperature life shortening in *Drosophila melanogaster*. *J. Gerontol.* **16**, 2–12.

STREHLER, B. L. (1961). Aging in coelenterates. *In* "Biology of Hydra and other Coelenterates" (H. M. Lenhoff and W. F. Loomis, eds.), Univ. Miami Press, Coral Gables, Florida.

STREHLER, B. L. (1962a). "Time, Cells, and Aging." Academic Press, New York.

STREHLER, B. L. (1962b). Further studies on the thermally induced aging of *Drosophila melanogaster*. *J. Gerontol.* **17**, 347–352 (1962).

STREHLER, B. L. (1963). The senescence of differentiated cells: Some chemical bases of cellular aging. *In* "General Physiology of Cell Specialization" (D. Mazia and A. Tyler, eds.) pp. 116–148. McGraw-Hill, New York.

STREHLER, B. L. (1964a). Studies on the comparative physiology of aging, III. Effects of X-radiation dosage on age-specific mortality rates of *Drosophila melanogaster* and *Companularia flexuosa*. *J. Gerontol.* **19**, 83–87.

STREHLER, B. L. (1964b). On the histochemistry and ultrastructure of age pigment. *In* "Advances in Gerontological Research" (B. L. Strehler, ed.), Vol. I, Academic Press, New York.

STREHLER, B. L. (1966a). Code degeneracy and the aging process: A molecular genetic theory of aging. *Proc. Intern. Congr. Geront., Vienna, 1966, Wien. Med. Acad.*

STREHLER, B. L. (1966b). Cellular aging: Fact and hypothesis. *Proc. S.E.B. Symp. Aging, Sheffield, England, 1966* (H. W. Woolhouse, ed.). Cambridge Univ. Press, Cambridge, England.

STREHLER, B. L. (1966c). Cellular aging. Symposium on Time, 1966. New York Academy of Sciences, New York.

STREHLER, B. L., AND MILDVAN, A. S. (1960). General theory of mortality and aging. *Science* **132**, 14–21.

STREHLER, B. L., MARK, D. D., MILDVAN, A. S., AND GEE, M. V. (1959). Rate and magnitude of age pigment accumulation in the human myocardium. *J. Gerontol.* **14**, 430–439.

STREHLER, B. L., KONIGSBERG, I., AND KELLY, J. (1963). Ploidy of myotube nuclei developing *in vitro* as determined with a recording double-beam microspectrophotometer. *Exptl. Cell Res.* **32**, 232–241.

STREHLER, B. L., WILDER, R. J., RAYCHAUDHURI, A., PRESS, G., AND GEE, M. (1966). Studies on the mechanism of cellular death. I. Enzymatic changes during early and late cardiac necrosis. *J. Exp. Gerontol.* **1**, 301–313.

STREHLER, B. L., WILDER, R. J., RAYCHAUDHURI, A., GEE, M., AND PRESS, G. (1967). Studies on the mechanism of cellular death. III. Changes in proteins and connective tissue elements during early and late cardiac necrosis. *J. Gerontol.* **22**, 52–58.

STREHLER, B. L., HENDLEY, D. D., AND HIRSCH, G. P. (1967a). Evidence on a codon restriction hypothesis of cellular differentiation. *Proc. Nat. Acad. Sci. USA* (in press).

STRYDOM, N. B. et al. (1963). Effect of heat on work performances in gold mines in South Africa. *Fed. Proc.* **22**, 893–896.

SZILARD, L. (1959). On the nature of the aging process. *Proc. Nat. Acad. Sci. USA* **45**, 30–45.

TEMIN, H. M. (1966). Genetic and possible biochemical mechanisms in viral carcinogenesis. *Cancer Res.* **26**, 212–216.

"United Nations Demographic Yearbook" (1955). 7th edition. Statistics Office, United Nations, New York.

UPTON, A. C. (1962). Age-specific death rates of mice exposed to ionizing radiation and radiomimetic agents. *In* "Cellular Basis and Aetiology of Late Somatic Effects of Ionizing Radiation" (R. J. C. Harris, ed.). Academic Press, New York.

VERZAR, F., AND THOENE, H. (1960). Die Wirkung von Elektrolyten auf die thermische Kontraktion von Collagenfaden. *Gerontologia* **4**, 112–119.

VON HAHN, H. P. (1964). Age-related alterations in the structure of DNA. II. The role of histones. *Gerontologia* **10**, 174.

VON HAHN, H. P. (1965). "Nucleinsauren und Nucleiproteine in Zellen alternder Organe." Habilitationsschrift, Basel.

WALFORD, R. L. (1966). The general immunology of aging. *In* "Advances in Gerontological Research" (B. L. Strehler, ed.), Vol. II, pp. 159–203. Academic Press, New York.

WEINBACH, E. C. (1959). Oxidative phosphorylation in mitochondria from aged rats. *J. Biol. Chem.* **234**, 412–417.

WEISMANN, A. (1891). "Essays Upon Heredity and Kindred Biology Problems." Oxford Univ. Press, London.

WILLIAMS, G. C. (1957). Pleiotropy, natural selection and the evolution of senescence. *Evolution* **11**, 398–411.

WULFF, V. J., QUASTLER, H., AND SHERMAN, F. G. (1962). An hypothesis concerning RNA metabolism and aging, *Proc. Nat. Acad. Sci. USA* **48**, 1373–1375.

WULFF, V. J., QUASTLER, H., AND SHERMAN, F. G. (1965). The incorporation of $H^3$-cytidine into some viscera and skeletal muscle of young and old mice. *J. Gerontol.* **20**, 34–40.

WULFF, V. J., SAMIS, H. V., JR., AND FALZONE, J. A., JR. (1966). The metabolism of ribonucleic acid in young and old rodents. *In* "Advances in Gerontological Research" (B. L. Strehler, ed.), Vol. II, pp. 37–75. Academic Press, New York.

## Commentary

### MAROTT F. SINEX

*Department of Biochemistry, Boston University, Boston, Massachusetts*

To evaluate the significance of external factors, it is necessary to have some understanding of the changes that result from internally produced noxious substances even in an optimal environment. Decrements in physiological function occur at a rate of about ten percent per decade. We are remarkably well equipped to survive deleterious environmental factors until we get older; as we get older, ability to cope decreases.

All would like better measures of what chemical changes occur in DNA with the passage of time in various environments. Chromosomal aberrations and point mutations must be measured separately. Most of the experiments done to prove the importance of somatic mutations for aging in effect disprove it. The most important inconsistency is between the profound effects that ionizing radiation has on chromosomal aberrations and the rather large amounts required to produce life shortening.

Little is known about the accuracy with which the phenotype is preserved in cellular division. Changes occur in the concentration of unusual amino acids in connective tissue cells with development and aging. The ability of these cells to make young connective tissue depends on the age of the animal. If one believes in environmental chemical injury, the lung is the place to look. More sensitive methods are needed for measuring injury to macromolecules, many of which do not undergo appreciable turnover.

# Ecologic and Ethnic Adaptations

J. A. HILDES

*Department of Medicine, University of Manitoba Medical College,
Winnipeg, Manitoba, Canada*

During the first sessions of this Symposium, the biochemists and pharmacologists gave us some insight into the effects of environmental hazards on men and on experimental animals. Other aspects of the study of health hazards of the environment have been indicated by Dr. Aschoff and by Dr. Prosser—namely, the adaptations which occur to environmental conditions. This is the area we are considering now.

We are primarily concerned with the ecosystem constituted by man in relation to his total environment. According to Dobzhansky (5), man's total environment comprises:
(1) the physical environment—temperature, humidity, etc.; (2) the biologic environment—the organisms which serve as food, as parasites, and as competitors; (3) the cultural or social environment—which man, amongst the other animals, has developed to a unique degree; and (4) the genetic environment—the genetic makeup of the individual.

We have heard about genetic adaptation. In addition to the capacity for evolutionary change and progress by genetic selection, mammalian organisms, including man, have a "phenotypic plasticity," which allows adaptations to occur within the life span of the individual (16). These may be changes of acclimation or acclimatization, as defined by Prosser. Some of these are reversible short-term effects such as the physiologic effects of muscular training, which may be induced at any phase of the life span. Other environmental stresses may induce certain adaptive responses only if they are operative in the early stages of development of the individual, and may be relatively irreversible.

Man's phenotypic plasticity extends to an area of such importance to him as to warrant special consideration—the adaptability of the central nervous system which is the basis for behavioral and cultural responses to the environment, that is, learning and social behavior (3). This type of adaptation is not only available to the individual, but it is capable of nongenetic transmission throughout the population and from generation to generation. Therefore, it can effect changes in a population very much faster than genetically dependent adaptation, as Dr. Li so emphatically pointed out.

To recapitulate, we have physiologic, relatively short-term adaptations, both acclimation and acclimatization; developmental adaptations; and cultural adaptations, by which individuals react to the physical, biological, and cultural aspects of their environment.

The cultural adaptations are those on which man chiefly relies for the very rapid adjustments necessitated by the considerable changes in the environment which he himself induces. Even allowing for our egocentricity, as Dr. Hardy brought out in the discussion, man is unique; he is the dominant animal in the world today, and through the control of energy he has wrought great environmental changes, including the elimination of other species and the development of some new ones. These transformations of the environment are usually undertaken without much thought of their ecologic consequences; in fact, some of the consequences may be unforeseeable.

Both the biologic and the cultural adaptations have genetically determined limitations, and may themselves influence genetic selections. Therefore, studies to separate out the various components of man's adaptive responses to his total environment may be expected to be somewhat complex. Nevertheless, our purpose at this Symposium is not only to outline some of the known environmental stresses and man's capacity for adaptive responses to them, but also to consider the types of investigation which may be needed to separate out the physiologic, cultural, and genetic components of the adaptive processes.

## ADAPTATION TO THE PHYSICAL ENVIRONMENT

There is abundant evidence that man has the capacity to adapt to many facets of the physical environment. Reference has been previously made at this Symposium to the "Handbook of Physiology," edited by Dill *et al.* (4), to which some of the participants at this Symposium have contributed.

Temperature and altitude are two of the environmental factors for which the evidence of adaptation is most abundant, and we will be hearing from Dr. Wyndham, later, on adaptations to heat and to cold. The same Handbook contained chapters outlining the state of knowledge with regard to solar radiation, light intensity, motion, and noise. We have heard from Dr. Aschoff about adaptive biologic rhythms. At the present time, there is little evidence of physiologic adaptation to ionizing radiations or to air pollution.

However, even for environmental temperature and barometric pressure, which have been areas of active interest and investigation for some years, our knowledge is rather fragmentary; some of the phenomena still require full description, mechanisms of action are not all clear, and the variability of individuals and populations are important aspects requiring study.

Without encroaching on Wyndham's ground too far, I will comment briefly on adaptations to environmental temperature to illustrate some of the points I wish to make. In a review of the studies of cold tolerance which have been made on primitive man, Hammel found the differences between the various ethnic groups impressive (7). He speculated that these differences were probably due to evolutionary progression based on genetic selection. However, Weiner (15), in discussing Hammel's paper, pointed out that, if this is so, it is very different from heat adaptation, for which no evidence of genetic differences between races exists.

The differences expounded by Dr. Prosser between acclimation and acclimatization are of importance in considering responses to cold. Adaptations to cold may not be direct and specific consequences of cold exposure *per se*. They are still "adaptive" if they confer an advantage on the individual or the population. For example, the dietary habits of Eskimos, with respect to the high intake of protein, may well be one basis for the increase in metabolism which confers a distinct advantage in cold exposure (14). Similarly, the recent studies by Andersen on Norwegian men support and extend other studies indicating that improved muscular ability, acquired through relatively short-term physical training, also confers a distinct advantage in the cold (2). In Andersen's study, it is of interest that concomitant cold exposure did not improve the tolerance to cold over that achieved by physical training alone.

Although man has occupied the Arctic and sub-Arctic regions (and now even the Antarctic) only through the development of technology and behavioral adaptation, the contribution to his comfort made by physiologic adaptations is still of interest and perhaps of some importance. Dr. Prosser expressed the view that the initial response to stress was often a behavioral one, followed later by physiologic adaptation, implying that the latter was more economical or more efficient for the organism. We would like to know the limits of the adaptive capacity imposed on various populations by their genetic background and the cost of acquiring and retaining physiologic adaptations.

Considering how to investigate these problems, it seems impractical to bring populations, or good samples of populations, to the laboratory for controlled studies and critical experiments. One has only to think of the time required to study human adaptations during the developmental period of an individual. We have to search for natural situations likely to answer the questions. Such a situation has been envisioned by Harrison *et al.* in a plan for identifying the components of adaptation to altitude in the Himalayan regions as part of the International Biological Program Human Adaptability Proposals (8). Essentially, the plan is a comparison of two populations, one indigenous to high altitudes but with recent migration to low altitude, and the other in the opposite situation. It is hoped that similar studies will be made on Arctic populations: genetically similar ones in different environments, and genetically dissimilar ones in the same environment. The examination of these populations must be intensive and include all aspects of their interaction with the total environment. It will not be sufficient to look merely at genetic markers and physiologic tests to heat, to cold, or to low barometric pressure. This lesson is clearly pointed out in the report by Edholm *et al.* of the performance of a military task in a hot climate (6). Of three groups of subjects, one was naturally acclimatized, one artificially acclimatized, and one nonacclimatized (control). Their ability to perform the military tasks in the heat bore little relationship to the physiologic parameters of adaptation to heat, but was probably more closely related to such uncontrolled factors as intelligence and motivation.

Similarly, in a study specifically designed to assess motivation in carrying out a particular task in a South African gold mine, Wyndham *et al.* (17) found that, although physiologic capacity imposed certain limitations on the amount of physical work performed during a work shift, the variability between individuals and the mean level of work performed were related more to the quality of the supervision than to either of the two physiologic parameters measured.

## BIOLOGIC FACTORS IN THE ENVIRONMENT

Two of the biologic factors in the environment to which both physiologic and cultural adaptations may occur are nutrition and infection. Man has successfully adapted to a great variety of foodstuffs and dietary patterns. Also, the amount of food energy available varies between populations and from time to time. It may be stretching a point to label the response to starvation and to chronic undernutrition as adaptive, but there is evidence that the organism reacts to these stresses in a way which promotes survival, and therefore may be considered adaptive in the sense. With respect to essential minerals, Mitchell (13) in his review of nutritional adaptations discussed the evidence of the adaptation of adult man to severe restrictions in dietary calcium. He concluded that some individuals have the capacity to adapt to low dietary calcium intake. However, McCance et al. in a recent review of the same topic consider it still to be an open question (11).

One aspect of the relationship between nutrition and cold tolerance that has interested me is the apparent dependence of the elevated basal metabolic rate of Eskimos on the dietary intake of protein. When two Caucasian men voluntarily used an exclusively meat diet for a year they did not develop a significant increase in basal metabolic rate (12). I wonder if the Eskimo has a genetically determined capacity for metabolizing larger amounts of protein than Europeans. One might look for differences in enzyme systems—in the gut for the absorption of protein, and in the body for metabolizing it. As far as I know there have been no studies of these points.

It has been said that infection has been one of the great selective forces in the evolution of man. The special advantages of heterozygosity with respect to sickle-cell trait in malarious areas is the example usually quoted (1). However, little is known of the tolerances developed to specific organisms by various populations. In an antibody survey carried out in conjunction with my colleagues at the University of Manitoba on Canadian Arctic Eskimos, we were rather surprised to find a remarkably high incidence of antibodies to psittacosis virus (10). At first, we were skeptical of the significance of these antibodies, but this has now been confirmed in a number of Eskimo and Arctic Indian populations, from different parts of the Canadian Arctic (9). We are not certain whether these antibodies represent the prevalence of a serious disease carrying a high mortality and morbidity which has gone unrecognized, or whether the relationship between the host and the parasite has been altered by some adaptive mechanism so that the virus has diminished virulence for these populations. Positive skin tests for trichinosis are also prevalent in the Arctic, but, since clinical cases of the acute disease with the usual severe features have been reported, this situation may be quite different.

## CULTURAL ASPECTS OF THE ENVIRONMENT

I have mentioned the cultural adaptations which man makes to the stresses in the environment, but we should also consider the special cultural aspects of the environment. These include social organization and attitudes, mating patterns, crowding, habitual level of physical activity, dietary habits, and invention and technology. Noise, automobiles, air pollution, drugs, clothing, and soap are all parts of our cultural environment. The habitual use of clothing may have interfered with our natural capacity to react to cold in a primitive way; and the automobile may greatly diminish the requirement for physical exertion. Hygiene may have greatly influenced not only the epidemiology of infectious diseases, but perhaps also our physiologic adaptations to the organisms.

## FINAL REMARKS

I think we may be confident that man will continue to change his environment at will, and at the same time decrease his adaptive requirements for the extremes of temperature, pressure, noise, physical fitness, nutritional deficiencies, etc. A major question is whether man is acquiring new adaptive capacities fast enough to keep up with the changes he is making in his environment; e.g., adaptations to air pollution, ionizing radiation, and crowding.

We cannot turn back the clock. Man is continuing to evolve, even though genetic evolution may no longer be the most important mechanism of this progressive change. We should now be turning some of our inventive genius away from considering ways to exploit our environment and toward the major problems of preserving our own health in a rapidly changing world.

## REFERENCES

1. ALLISON, A. C. (1954). Protection afforded by sickle-cell trait against subtertian malarial infection. *Brit. Med. J.* 1, 290–294.
2. ANDERSEN, K. L. (1966). Metabolic and circulatory aspects of tolerance to cold as affected by physical training. *Federation Proc.* 25, 1351–1356.
3. BAKER, P. T. (1965). "International Biological Programme. Multidisciplinary Studies of Human Adaptability; Theoretical Justification and Method." H. A. 60. Special Committee for the International Biological Programme, 7 Marylebone Road, London, N.W.1. by Giannini, Naples, pages 63–71.
4. DILL, D. B., ADOLPH, E. F., AND WILBER, C. G. eds. (1964). "Handbook of Physiology," Sect. 4. Am. Physiol. Soc., Washington, D. C.
5. DOBZHANSKY, T. (1962). "Mankind Evolving. The Evolution of the Human Species." Yale Univ. Press, New Haven, Connecticut.
6. EDHOLM, O. G., FOX, R. H., ADAM, J. M., AND GOLDSMITH, R. (1963). Comparison of artificial and natural acclimatization. *Federation Proc.* 22, 709–715.
7. HAMMEL, H. T. (1963) Summary of thermal patterns in man. *Federation Proc.* 22, 846–847.
8. HARRISON, G. A., BAKER, P. T., AND WEINER, J. S. (1966). "Biology of Human Adaptability." p. 509. Clarendon Press, Oxford.
9. HILDES, J. A. (1966). A note on psittacosis antibodies in Canadian Eskimos and Arctic Indians. *Intern. Pathol.* 7, 74–75.
10. HILDES, J. A., WILT, J. C., AND STANFIELD, F. J. (1958). Antibodies to adenovirus and to psittacosis in Eastern Arctic Eskimos. *Can. J. Public Health* 49, 230–231.
11. MCCANCE, R. A., WIDDOWSON, E. M., EDHOLM, O. G., AND BACKARACH, A. L. (1965). "The Physiology of Human Survival," pp. 207–233. Academic Press, New York.
12. MCCLELLAN, W. S., SPENCER, H. J., AND FALK, E. A. (1931). Clinical calorimetry: prolonged meat diets with a study of respiratory metabolism. *J. Biol. Chem.* 93, 419–434.
13. MITCHELL, H. H. (1964). *In* "Nutrition, A Comprehensive Treatise" (G. H. Beaton, and E. W. McHenry, eds.) Vol. 2, p. 351. Academic Press, New York.

14. RODAHL, K. (1962). Basal metabolism of the Eskimo. *J. Nutr.* **48**, 359–368.
15. WEINER, J. S. (1963). Discussion. *Federation Proc.* **22**, 849.
16. WEINER, J. S. (1964). *In* G. A. Harrison, J. S. Weiner, J. M. Tanner, and N. A. Barnicot. "Human Biology." Clarendon Press, Oxford.
17. WYNDHAM, C. H., COOKE, H., MUNRO, A., AND MARJTℓ, J. (1964). The contribution of physiological factors to the performance of moderately heavy physical work. *Ergonomics* **7**, 121–137.

## Commentary

### CYRIL H. WYNDHAM

*Human Sciences Laboratory, Chamber of Mines, Johannesburg, South Africa*

It is easy to fall into the trap of attributing physiological differences to genetic factors which, in fact, are the result of differences in nongenetic biological and cultural aspects of the environment. The Kalahari bushmen, for example, have not had more than a hundred years for adaptation and in objective testing show the same response as the Bantu. Their adaptation to cold is cultural, through modification of their microenvironment so that they are not subjected to the cold stress that the macroenvironment might suggest.

There has been inadequate attention to the possibility that the effects of drugs or toxic substances on experimental groups may be modified by differences in nutrition, activity patterns, or the climate in which the animals are maintained. Prosser emphasized the necessity for standardizing extremely carefully a whole series of environmental factors before drawing conclusions from metabolic studies on animals. The selection of animals for experimental work seems to have been made more on their availability than on the significance of their reactions for the effects of exposure in man.

# Ecological Implications of Individuality in the Context of the Concept of Adaptive Strategy[1]

FREDERICK SARGENT

*College of Envivonmental Sciences,
University of Wisconsin—Green Bay, Green Bay, Wisconsin*

Strategic planning to meet environmental problems affecting man involves three fundamental elements: the biological population, the natural ecosystem, and the behavioral codes by which conditions are controlled. An important component of the intelligence on which strategy can be based is Dubos' "prospective epidemiology": (1) systematic efforts to determine beforehand the risks involved in any social and technological innovation; and (2) detailed and continued surveillance of the population at risk.

Health must be defined and measured in terms of the adaptive capacity of man towards environmental circumstances and hazards. Health is successful adaptation; illness is lack of success. Ill-health will never be eradicated; it can be reduced. With all the spectacular successes in medical science, there has been no essential change in the potential life span. More individuals get to utilize more of the potential span, but the ultimate point remains unchanged.

There has been an extraordinary acceleration in the rate of environmental change. As much major change takes place now in one year as required about 180 years 8 to 10 thousand years ago. The change is man-made; it can be reviewed and corrected by man.

Man is an organism with a broad range of adaptability through both genetic diversity and phenotypic plasticity. But the adaptability is limited. That which is due to selection is past-oriented. The reciprocal fitness of man and his environment can be disturbed. On the other hand, not all would be equally affected; groups differ in the qualities of their optimum environment. In respect to some environmental conditions, such as the climatic factors, man has a fairly wide range of adaptability; to others, such as water deprivation, he has little.

The socio-cultural factors present problems. In the first place, their increasing complexity tends to impose increasing restraint. On the other hand, since they are responsible for much of the current environmental difficulty, they can be modified to help. The pertinent questions are: (1) Can man adapt to a polluted environment? (2) Can he adapt to great constraints on his activities? (3) Can he adapt to the increasingly diverse demands made upon his time and loyalties? and, most fundamental, (4) can he raise his mode of thinking and working from the parochial level of in-group values to the level of mankind as a whole?

Each individual must simultaneously play several games of survival. Since both his consumption capacities and life are limited, his optimal tactics are

[1] Synopsis of presentation, published in *Intern. J. Biometeorol.* **10**, 305–324 (1966).

directed towards maximizing his security level rather than pursuing maximum gain. Mixed tactics are most powerful in the situations of incomplete information that characterize most real events.

The "constants" of bodily processes show a great variability, as befits their probably polygenetic nature. Individuality is usually exemplified by a particular constellation of values representing these "constants." Most individuals exhibit specific departures from the boundaries of normal growth channels. Charts of physico-chemical levels of internal environmental conditions exhibit great individuality, as do measures of homeostatic processes. The mean calorie consumption of a group living under common conditions reveals great individuality. If 100 traits are taken into consideration, then only 0.59 percent of a population can be expected to be normal with respect to all traits. Equal homeostasis can be obtained by different configurations of regulatory processes.

Polymorphisms are expressions of genetic diversity within a gene pool containing different alleles of many genes. They provide insurance against the future. A population of human beings consists of individuals who may be arrayed according to configurations of constants into various modal subsets or biotypes. Each biotype has characteristics and distinctive adequate environments, needs and requirements. In defining biotypes, care has to be taken to distinguish those individuals who are truly sensitive to a given factor from those who react by suggestion. There is a certain complex of complaints that can be called the "Flying Dutchman syndrome"—they never die but keep coming back attached to whatever is the fashionable "cause" at the moment.

The central ecological problem posed by prospective epidemiology is the measurement of risk. Risk must be assessed in terms of the individual's characteristics and what constitutes his adequate environment. This provides the desideratum of positive health, which varies with the biotype. Definition of the biotypes would identify the groups particularly susceptible to an environmental factor and improve the precision of epidemiological studies. A feasible battery of clinical field procedures is required to identify the individual in terms of mathematically independent traits. The investigations must be conducted longitudinally on representative stratified cohorts of the population, so that changes with time can be measured.

Human populations must give their consent and release for this type of study. If he only knew it, the individual is already at risk; he is already submitting himself unwittingly to a variety of major manipulations of the ecosystem. Man's attitude toward human examination will have to change from the parochial indifference now widespread to one that appreciates the concept of an adaptive strategy for mankind's survival.

## Commentary

A. PHARO GAGGE

*John B. Pierce Foundation Laboratory, Yale University,
New Haven, Connecticut*

and

GEORGE Z. WILLIAMS

*Clinical Center, National Institutes of Health,
Bethesda, Maryland*

In studies of human reactions to heat, in which individuals adjusted room temperature to maintain comfort in the face of imposed variations in radiant heat, each subject was found to have his own but consistent idea of the combination he identified as comfortable. Attempts to correlate the individuality with various physiological functions was without success, although there was some indication that the subject's internal body temperature may serve as a reference.

Measurements of the levels of chemical constituents in blood from fasting subjects showed a relatively narrow range in each of 29 subjects as compared with the normal range, and differences in this range between subjects. There was little change in the individual profiles over 2 years, but undoubtedly changes will occur with age. More precise methods must be developed for many laboratory measurements before they can be used to study the normal variance in a population. Base-line profiles could prove very useful, not only in epidemiology, but also in health management of the individual. Education will be required of the public, industry, and physicians if this new tool is to be utilized.

# Cross-Adaptation

## Henry B. Hale

*School of Aerospace Medicine, Brooks Air Force Base, San Antonio, Texas*

Cross-adaptation is an aspect of environmental physiology which has not been explored extensively. The knowledge on cross-adaptation is fragmentary, and no generalizations are currently possible. The concept of cross-adaptation is quite simple, merely holding that long-term exposure (either continuous or intermittent) to a given adverse environment not only leads to an increase in tolerance to that particular environment but leads also to gains or losses in tolerance to other adverse factors, ones which the adapted organism had never encountered previously. In other words, for any given adaptation, some side effects are to be expected, and these effects may be either good or bad. High degrees of physiologic displacement resulting from exposure to a test environment are taken as evidence of "crossed sensitization" or "negative cross-adaptation," while low degrees of displacement under such conditions denote "crossed resistance" or "positive cross-adaptation."

Obviously, cross-adaptations of many types may occur, since the factors that constitute the environments of man are numerous and differ greatly in their nature. Programs of research on cross-adaptation that might be generated or promoted by the Committee on Environmental Physiology, therefore, should be very broad and deal with many different factors, each considered from a spectral point of view. Temperature and pressure spectra, for example, have been utilized in adaptation and cross-adaptation studies. In fact, the few published papers that have been expressly concerned with cross-adaptation (5, 15, 19, 26–36, 38, 40, 41, 50, 57, 58, 74) dealt with the factors of low barometric pressure, cold, or heat. The results obtained in some of these studies will be discussed briefly later, since they show that cross-adaptations are highly variable. While adaptational patterns of change have been worked out for a great number of physiologic functions, cross-adaptation patterns and principles remain to be determined.

Studies on cross-adaptation should deal not only with single-agent environments, but also with multiagent environments; and numerous environmental and nonenvironmental agents should be studied.

In each case, the factors of intensity and duration must be examined thoroughly. When the intensity factor is low, it is often difficult to demonstrate physiologic influence; yet the adaptation process may have been set into motion, and cross-adaptation tests may provide evidence of such change. In many studies of adaptation, a particular physiologic activity may become elevated greatly at the beginning of the period of adaptation and subsequently show gradual decline to the pre-exposure level. If the adapted organism is taken from the experimental environment and tested under the condition in which control animals were kept, an undershoot may appear, thereby indicating that adaptive changes had existed. Further testing in other adverse environments, then, should reveal cross-adaptation. Many metabolic and endocrine functions in altitude-acclimated animals tend to follow the pattern of initial overshoot and gradual return to the preexposure level. Particularly pertinent are the findings of Fregly (33), who observed negative cross-adaptation both between cold and altitude and between altitude and cold. It was in body-temperature regulation that the carryover of the antecedent adaptations became evident. The extensive work of Flückiger and Verzar (27–32) indicates pituitary-thyroid-adrenal involvement in cross-adaptation; these investigators also studied the negative cross-adaptation between altitude and cold. Fleischner and Sargent (26) found negative cross-adaptation between heat and cold. Particularly important was their finding that the carryover effects of either heat or cold acclimation were not readily reversible. Human studies by Davis (19), Glaser and Shephard (34–36) and Stein, Eliot, and Bader (74) also show persistency of adaptations; but Davis concluded that heat and cold acclimatizations are not mutually exclusive and can exist simultaneously in man. The observations of the other authors also support the conclusion that "deadaptation" is not a necessary step in sequential adaptations. It would be worth while to determine whether deadaptation proceed at different rates in different environments and if factors in combination accelerate adaptation or deadaptation.

The results obtained in cross-adaptation studies performed at the USAF School of Aerospace Medicine (38, 40, 41, 57, 58) show that negative and positive cross-adaptations can coexist. Illustrative results are given in Tables I–III. Table I presents 24-hr urinary excretion rates for nitrogenous metabolites and electrolytes of rats adapted (acclimated) to cold (5°C), to laboratory temperature (25°C), or to heat (35°C). The period of acclimation was 3 months, which is adequate to bring acclimation to an advanced state. To test for cross-adaptation, 20 of the animals from each thermal background were exposed for 24 hr to low barometric pressure (380 mm Hg, with ambient temperature equal to

TABLE 1
Cross Adaptation
(Altitude test)

| Urinary variable[a] | Cold-adapted (n = 20) | | Laboratory-adapted (n = 20) | | Heat-adapted (n = 20) | | Altitude-adapted (n = 20) |
|---|---|---|---|---|---|---|---|
| (Weight loss, g) | *72* | 63 | *42* | 57 | *50* | 54 | 47 |
| Creatine, mg | *11.6* | 9.9 | *2.6* | 6.2 | *2.4* | 4.7 | 2.2 |
| Urea, mg | *1410* | 1344 | *1066* | 1159 | *1013* | 1172 | 1039 |
| Uric acid, mg | *4.5* | 5.8 | *5.2* | 7.5 | *4.8* | 5.4 | 4.5 |
| Phosphorus, mg | *51* | 45 | *36* | 39 | *29* | 31 | 34 |
| Magnesium, mEq | *0.86* | 1.42 | *0.83* | 1.17 | *0.85* | 0.96 | 0.74 |
| Potassium, mEq | *3.50* | 3.82 | *1.98* | 2.88 | *2.12* | 3.14 | 2.40 |
| Sodium, mEq | *0.77* | 1.69 | *0.89* | 1.59 | *1.25* | 1.29 | 0.78 |

[a] Quantity/24 hr/kg$^{.75}$
Values for unexposed controls are given in italics.

## TABLE 2
### Cross Adaptation
### (Heat test)

| Urinary variable[a] | Laboratory-adapted (n = 20) | | Altitude-adapted (n = 20) | | Cold-adapted (n = 20) | | Heat-adapted (n = 20) |
|---|---|---|---|---|---|---|---|
| (Weight loss, g) | *42* | 71 | *47* | 50 | *72* | 50 | 50 |
| Creatine, mg | *2.6* | 3.6 | *2.2* | 5.4 | *11.6* | 4.8 | 2.4 |
| Urea, mg | *1066* | 1227 | *1039* | 1165 | *1410* | 1323 | 1013 |
| Uric acid, mg | *5.2* | 6.6 | *4.5* | 4.7 | *4.5* | 5.4 | 4.8 |
| Phosphorus, mg | *36* | 43 | *34* | 70 | *51* | 36 | 29 |
| Magnesium, mEq | *0.83* | 0.74 | *0.74* | 0.73 | *0.86* | 1.21 | 0.85 |
| Potassium, mEq | *1.98* | 1.92 | *2.40* | 1.90 | *3.50* | 3.04 | 2.12 |
| Sodium, mEq | *0.89* | 1.82 | *0.78* | 0.90 | *0.77* | 0.60 | 1.25 |

[a] Quantity/24 hr/kg$^{.75}$
Values for unexposed controls are given in italics.

that in the laboratory). Food was withheld during the test period; consequently, body-weight losses occurred. Additional animals from each of the adapted groups were tested while remaining in the environments to which they were adapted; they were also deprived of food. The smaller numerals in the table give the values for the animals that remained in their accustomed environments; and the larger numerals give values for the acutely altitude-exposed animals. Altitude effects in each group of adapted animals, therefore, can be quantified. In addition, each group of acutely altitude-exposed animals has been compared with altitude-adapted animals, ones that had remained for 3 months at 380 mm Hg barometric pressure and then studied while deprived of food for 24 hr and remaining in the accustomed environment.

Weight losses for the three acutely altitude-exposed groups exceeded that of the altitude-adapted group, just as would be expected. The fact that the different thermal backgrounds induced different degrees of sensitivity to altitude would not necessarily be expected. The lower body-weight loss of the

## TABLE 3
### Cross Adaptation
### (Cold test)

| Urinary variable[a] | Heat-adapted (n = 20) | | Altitude-adapted (n = 20) | | Laboratory-adapted (n = 20) | | Cold-adapted (n = 20) |
|---|---|---|---|---|---|---|---|
| (Weight loss, g) | *50* | 40 | *47* | 58 | *42* | 66 | 72 |
| Creatine, mg | *2.4* | 12.1 | *2.2* | 5.6 | *2.6* | 11.6 | 11.6 |
| Urea, mg | *1013* | 1404 | *1039* | 1553 | *1066* | 1579 | 1410 |
| Uric acid, mg | *4.8* | 5.1 | *4.5* | 4.2 | *5.2* | 6.5 | 4.5 |
| Phosphorus, mg | *29* | 76 | *34* | 87 | *36* | 61 | 51 |
| Magnesium, mEq | *0.85* | 0.91 | *0.74* | 0.95 | *0.83* | 1.18 | 0.86 |
| Potassium, mEq | *2.12* | 3.98 | *2.40* | 3.44 | *1.98* | 3.56 | 3.50 |
| Sodium, mEq | *1.25* | 0.49 | *0.78* | 0.50 | *0.89* | 1.15 | 0.77 |

[a] Quantity/24 hr/kg$^{.75}$
Values for unexposed controls are given in italics.

heat-adapted group indicates that these animals had a slight advantage over the laboratory-adapted ones and definite superiority over the cold-adapted ones. Further support to this conclusion is provided by their excretion, of creatine, uric acid, phosphorus, magnesium, and sodium, since in all of these respects the altitude-exposed, heat-acclimated animals were below the levels found for the two other groups. Uniquely, phosphorus excretion for the exposed heat-adapted group was even lower than that for the altitude-adapted group itself, thereby suggesting an unusually high degree of positive cross-adaptation.

The thermal backgrounds had numerous carryover effects; still, close inspection of these data will show that altitude effects and thermal effects tend to be intermingled, and the resulting effects are ones that could not have been predicted. Particularly interesting is the finding that cold-adapted animals, when acutely exposed to altitude at room temperature, experienced less weight loss than was observed in the accustomed environment; nevertheless, the values for uric acid, magnesium, potassium, and sodium in the altitude-exposed, cold-adapted rats exceeded those obtained in cold. These findings lead to the conclusion that negative and positive cross-adaptations can exist simultaneously.

Table II presents data for laboratory-adapted, altitude-adapted, and cold-adapted animals subjected to acute heat exposure. Test duration was 24 hr, and the animals, of course, were deprived of food. On the basis of body-weight losses, the altitude- and cold-adapted animals appear to be positively cross-adapted, showing no more weight loss than was found in heat-adapted animals; but, surprisingly, the laboratory-adapted animals exhibit negative cross-adaptation, judging by the relatively high values for weight loss, uric acid excretion, and sodium excretion. On the basis of weight loss, uric acid excretion, and sodium excretion, the altitude-adapted group shows positive cross-adaptation. Positive cross-adaptation is also evident for the altitude-adapted group in magnesium and potassium responses, but negative cross-adaptation is indicated by creatine, urea, and phosphorus responses.

Table III presents the results obtained for heat-adapted, altitude-adapted and laboratory-adapted animals subjected to acute cold exposure. As was noted in Tables I and II, the differences in background clearly altered the responses of the different groups; and there were differential effects among the selected physiologic functions.

Large batteries of measurements must be used in cross-adaptation studies in order to appraise broadly the physiologic state both in the accustomed environment and in test environments. The very extensive exploration of Sargent and Johnson (68–72), while not designed as a cross-adaptation study, provides a fine example of a multiparameter study. Their battery included many physiologic functions (endocrine, neurologic, metabolic, body composition, hepatic, renal, gastrointestinal, etc.); and they summarized their findings, as extensive as they were, in the form of nomograms which are easily read and which should have great usefulness.

Interrelations of functions, rather than the separate functions, appear to be of special importance in adaptations. Multiple correlational techniques, therefore, are appropriate for bringing the many phases of an adaptational response into proper perspective; and through such analysis it should be possible to develop what might be called an "adaptation index," one in which physiologic, psychologic, immunologic, or even other aspects can be treated in a collective fashion. Once derived, this index could be utilized in interdisciplinary research efforts. Such an index would probably be especially useful in military, industrial, and sports medicine. Dr. Roy B. Mefferd, who is Director of the Psychiatric and Psychosomatic Research Laboratory in the Veterans' Administration Hospital in Houston, is currently developing just such an index. He has tested it on various groups and under various circumstances. In a test performed under USAF contract on 122 volunteer subjects, he made 102 measurements, which included electrolytes in saliva, plasma, and urine; 17-hydroxycorticosteroids in plasma and urine; urinary nitrogenous metabolites (including a number of amino acids); phenolic-indolic acids in urine; catecholamines in urine; seven paper-pencil, perceptual-abilities tests; blood pressure; heart rate; oral temperature; and salivary and urine flow rates. The subjects were studied over a normal night's sleep, as well as immediately before, during, and after exposure to a simulated altitude of 15,000 ft. Multivariate analyses showed that certain of the functions under study remained correlated at all times, although altitude effects on the separate functions were clearly demonstrable. In certain other respects, altitude caused dissociations. Particularly important was the finding that interindividual differences could also be dealt with easily.

Trumbull (8), in a literature survey concerned with environmental modification of human performance, emphasized the interaction between the physical, social, cultural, and personal environments. Physiologic adaptation and cross-adaptation were areas that he felt needed much more study. He cited Balke (5), who reported improved capacity for work in hot climates in men who had been acclimatized to cold plus altitude. In addition, he cited Brüner et al. (15), who reported that altitude acclimatization improved human tolerance to heat, cold, and acceleration; and he also cited Jovy et al. (50), who found that 4 weeks of intermittent exposure to a simulated altitude of 6500 m gave men as much gain in altitude tolerance as resulted from 4 weeks of intermittent exposure to heat. Altitude tolerance was quantified by means of a psychomotor measure. Hiestand et al. (44) observed, in animal experiments, that there was increased tolerance to hypoxia after intermittent heat adaptation. Bartlett and Phillips (6) found that, after adaptation to a still different circumstances, long-term confinement, altitude tolerance of rats was increased to a considerable degree.

Findikyan and Sells (25), in reviewing the literature on cold stress, concerned themselves with the interaction of heat and cold with humidity, air movement, clothing insulation, and physical exercise and the resulting effect on the psychophysiologic behavior of the human organism, pointing out that the facilitating or inhibiting effects on resistance to thermal stress of air ions, radiation, drugs, diet, isolation, fear, darkness, motivation, personality traits, group structure, task orientation, and task load have not been adequately investigated. They made the point that, while a thermal factor is treated as a single influence, the additive or synergistic effects (or even the antagonistic effects) of other simultaneously applied stressors must be appraised carefully, since the presence of multiple stressors tend to lower both the physiologic and the psychologic resistance of the organism.

In the writings of Selye (73), crossed resistance (nonspecific resistance) and crossed sensitization are mentioned frequently. As a speculation, he suggested that adaptability (nonspecific resistance) falls whenever an organism becomes highly adapted to a particular environment. He has also suggested that positive cross-adaptation may develop when the adapting factor is of such strength as to induce mild effects, whereas very intense stimulation may induce negative cross-adaptation. These possibilities remain to be proved. He also indicated that positive cross-adaptation may be limited to early phases of the adaptation process.

The influence of climatic factors on sensitivity to allergens, to bacteria and their toxins, or to chemical agents (pharmacologic or nonpharmacologic) requires additional investigation. Unusual sensitivities to any of the above-mentioned agents may represent negative cross-adaptations. Conversely, regular use of pharmacologic agents (antihistamines, sympathomimetic or sympatholytic drugs, tranquilizers, diuretics, antibiotics, hormones, etc.) may have cross-adaptive influence toward climatic or other environmental factors, as may such other commonly used agents as tobacco, alcohol, and caffeine, especially when there is heavy use. The cross-adaptation concept can be extended to include occupations, exercise, and dietary deficiencies of excesses. Selye (73) cited authors who reported that (1) intense muscular effort led to decreased tolerance to cold; (2) undernourishment reduced tolerance to infection; and (3) starvation increased resistance to either hypoxia or hyperoxia. Burn (16) discussed the effects of changes in body temperature on drug actions and rates of drug degradation, but the possibility that adaptive states also affect drug actions was not considered. Drug dosages that are ordinarily quite satisfactory may be too high or too low for persons who are undergoing adaptation of one form or another. In discussing drug-induced diseases, Meyler and Peck (60) emphasized long-term accumulations of drugs as one cause of unusual sensitivity. Adaptation may be an augmenting factor. Environmental factors may affect drug tolerance in many ways; generally they do so by affecting metabolic, neuroendocrine, hepatic, or renal activities; and feedbacks complicate the picture greatly. Interaction of drugs and environmental factors may take place at various sites, ranging from regulatory mechanisms to target tissues. Drugs that mimic the actions of a given environmental factor may be useful for preadapting persons or for accelerating, augmenting, or reinforcing adaptations. Currently, Drs. Cain and Kronenberg at the USAF School of Aerospace Medicine are testing Diamox for its capacity to accelerate altitude acclimatization. Through the use of appropriate drugs, the adverse effects of environments may be circumvented, without affecting the beneficial ones.

According the Meyler and Peck (60), various drugs predispose infection, some acting nonspecifically, which possibly means that they induce negative cross-adaptation. Berry and Mitchell (8–11) have investigated extensively the influence of pressure and temperature on resistance to infection. Miya et al. (61) have studied specific and nonspecific resistance in relation to environmental temperature, as has Trapani (80). The interactions of drugs and hormones with environmental factors have interested numerous investigators (7, 20–24, 45, 49, 51, 54, 59, 78, 82, 83).

The programmed research of the U. S. Army Research Institute of Environmental Medicine, Natick, Massachusetts, now includes or will include studies designed to give information on adaptive changes, cross-adaptation, and drug-environment interactions.

Currently, there is widespread interest in the physiology of weightlessness. Attempts to simulate weightlessness or to diminish to some extent the influence of gravity have now become numerous. Deadaptation occurs during prolonged bedrest or water immersion, two conditions of reduced gravity; consequently, orthostatic intolerance develops. Since the cardiovascular changes during bedrest are opposite to those during altitude adaptation, Lamb (53) considered using hypoxia for the purpose of counteracting the deadaptation changes in bedrest. Cooper and Leverett (18), in an attempt to increase g-tolerance, used 3 months of vigorous exercise, but did not find evidence of either positive or negative cross-adaptation.

Endocrine and metabolic involvement in the adaptation process is well-established. Hormone interactions, rather than the separate effects of hormones, need to be studied in relation to adaptation and cross-adaptation. Particularly prominent are the catecholamines, along with pituitary, thyroid, and adrenocortical hormones, Chatonnet (17) has discussed the sites at which these hormones interact and their collective influence during adaptation to cold.

Although there has been interest in the possible differences between adaptations accomplished by means of continuous exposure and those resulting for intermittent exposure (2–4, 37, 48, 65, 75), the possibility that continuous and intermittent treatments alter cross-adaptations seems not to have been investigated extensively, if at all. The research on the effects of intermittent exposure of humans to altitude seems to be limited to the efforts of Hurtado et al. (48). Lind and Bass (56) lead in the research on intermittently heat-exposed humans, but they have not reported on cross-adaptations in intermittently heat-exposed persons.

The usual criterion of adaptation is a diminished physiologic response to the agent of interest. This diminished responsiveness may be due to diminished sensitivity in sensory mechanisms, changes in synaptic activity in the ascending or descending neural systems, changes within the central nervous system, changes in the endocrine system, or diminished end-organ capability. Because of feedback possibilities, the neural and endocrine changes may be quite complex. Hepatic transformations of hormones may change greatly during adaptations, and therefore be factors in cross-adaptations. Berry and co-workers (12–14, 46, 47) have examined metabolic functions during altitude adaptation and their relations to infection. Altitude adaptation has considerable influence on enzyme systems (64, 66, 67, 76, 77). Although altitude residents have normal plasma levels for 17-hydroxycorticosteroids and catecholamines, they may exhibit insulin hypersensitivity (62).

Considerable evidence has accumulated which indicates endocrine involvement in heat adaptation (39, 42, 43, 52, 55, 63, 79, 84–86). An extensive literature on endocrine changes during cold adaptation has accumulated also. Cold adaptations are better known than heat adaptations. The possible relations of the endocrine changes in either heat or cold adaptation to cross-adaptations remain to be explored. Some of the seasonal disease propensities may represent cross-adaptations. Seasonal variation in interrelated endocrine-metabolic activities and the intercorrelations with immunologic changes merits study. Persons with outdoor and indoor occupations should be compared, as should dayworkers and nightworkers and persons who are required to change shifts from time to time. Persons who experience time displacement may show some unusual reactions, since their diurnal trends may be altered. For example, flying personnel (either military or civilian) who make transcontinental or intercontinental flights and drivers of transcontinental trucks and buses are ones whose diurnal trends may show adaptational or cross-adaptational peculiarities.

Since the signs of human adaptation may not be readily detected, it is important at the outset to utilize large batteries of measures and to employ multivariate analyses. When possible, field testing should be performed, with the subjects following everyday routines. Urinalysis seems the most convenient means for appraising, in a multivariate manner, large segments of the population.

In summary, in this discussion I have emphasized that cross-adaptation is an aspect of environmental physiology that has been studied inadequately. The available literature on cross-adaptation, although meager, shows that research in this area will be fruitful. Many combinations of antecedent and test factors can be studied from the cross-adaptation viewpoint. Physiologists, pharmacologists, biochemists, psychologists, allergists, immunologists, and others should find this an area in which to combine their efforts.

## REFERENCES

1. ALBAUM, H. G., AND CHINN, H. I. Enzymes studies in acclimatization to high altitudes. I. Brain metabolism during altitude acclimatization. USAF School of Aviation Medicine Project No. 21-1201-0009, Report No. 1, May 1953.
2. ALTLAND, P. D. (1948). Acclimatization response of the adrenal gland of rats during discontinuous exposures to high altitude. *Proc. Penn. Acad. Sci.* **22**, 35.
3. ALTLAND, P. D., AND HIGHMAN, B. (1951). Acclimatization response of rats to discontinuous exposure to simulated high altitudes. *Am. J. Physiol.* **167**, 261.
4. ALTLAND, P. D., AND HIGHMAN, B. (1952). Effect of repeated acute exposure to high altitude on longevity in rats. *Am. J. Physiol.* **168**, 345.
5. BALKE, B. (1959). Experimental studies on the conditioning of man for space flight. *Air Univ. Quart. Rev.* **11**, 61.
6. BARTLETT, R. G., JR., AND PHILLIPS, N. E. (1960). Restraint adaptation and altitude tolerance in the rat. *J. Appl. Physiol.* **15**, 921.
7. BASS, D. E., AND JACOBSON, E. D. (1965). Effects of salicylates on acclimatization to work in the heat. *J. Appl. Physiol.* **20**, 70.
8. BERRY, L. J. (1956). Susceptibility to infection as influenced by acclimatization to altitude and Krebs cycle inhibitors. *J. Infect. Diseases* **98**, 21.
9. BERRY, L. J. The effect of environmental temperature on lethality of endotoxin and its effect on body temperature in mice. Arctic Aeromedical Laboratory Technical Report 65-12, December 1965.
10. BERRY, L. J. Tissue citric acid content and susceptibility to infection in mice acclimatiz-

ing to and recovering from altitude. USAF School of Aviation Medicine Report No. 56-110, August 1956.
11. BERRY, L. J., AND MITCHELL, R. B. Influence of acclimatization to altitude on susceptibility to intrastomachal infection with Salmonella typhimurium. USAF School of Aviation Medicine, Project No. 21-35-005, Report No. 7, November 1952.
12. BERRY, L. J., AND MITCHELL, R. B. The influence of polycythemia produced at high altitude on resistance to infection. USAF School of Aviation Medicine Project No. 21-35-005, Report No. 1, February 1951.
13. BERRY, L. J., AND SMYTHE, D. S. Effect of cortisone on protein loss and carbohydrate gain in normal and altitude-exposed mice. USAF School of Aviation Medicine Report 60-89, September 1960.
14. BERRY, L. J., AND SMYTHE, D. S. Effect of pure oxygen at reduced pressures on metabolic changes in mice living under simulated biosatellite conditions. USAF School of Aviation Medicine Report 62-24, January 1962.
15. BRÜNER, H., JOVY, D., AND KLEIN, K. E. (1961). Hypoxia as a stressor. *Aerospace Med.* **32**, 1009.
16. BURN, J. H. (1961). The effect of temperature on the response to drugs. *Brit. Med. Bull.* **17**, 66.
17. CHATONNET, J. (1963). Nervous control of metabolism. *Federation Proc.* **22**, 729.
18. COOPER, K. H., AND LEVERETT, S., JR. (1966). Physical conditioning versus $+G_z$ tolerance. *Aerospace Med.* **37**, 462.
19. DAVIS, T. R. A. (1961). The effect of heat acclimatization on artificial and natural cold acclimatization in man. U. S. Army Medical Research Laboratory Report No. 495. 1961.
20. DEBIAS, D. A. (1962). Hormonal factors in the rat's tolerance to altitude. *Am. J. Physiol.* **203**, 818.
21. DEBIAS, D. A., PASCHKIS, K. E., AND CANTAROW, A. (1958). Effects of chlorpromazine and autonomic nervous system blocking agents in combating heat stress. *Am. J. Physiol.* **193**, 553.
22. DEBIAS, D. A., PASCHKIS, K. E., CANTAROW, A., AND FRIEDLER, G. (1957). The effects of various metabolites and autonomic blocking agents in combating stress. *Exptl. Med. Surg.* **15**, 30.
23. DEWEY, M. L., AND LEUNG, S. E. C. Research on the influence of variations in environmental temperatures on the systemic effects of alcohol alone and in combination with other drugs. Arctic Aeromedical Laboratory Report AAL-TR-65-2, October 1965.
24. EVONUK, E., AND HANNON, J. P. (1963). Pulmonary function during norepinephrine-induced calorigenesis in cold-acclimatized rats. *J. Appl. Physiol.* **18**, 1213.
25. FINDIKYAN, N., AND SELLS, S. B. Cold stress: Parameters, effects, mitigation. Arctic Aeromedical Laboratory Technical Report 65-5, September 1965.
26. FLEISCHNER, J. R., AND SARGENT, F. II. (1959). Effects of heat and cold on the albino rat: crossed resistance or crossed sensitization. *J. Appl. Physiol.* **14**, 789.
27. FLÜCKIGER, E. Der Sauerstoffverbrauch der Ratte bei vermindertem Sauerstoffpartialdruck. *Helv. Physiol. Pharmacol. Acta* **14**, 369.
28. FLÜCKIGER, E. (1953). Störung der Thermoregulation nach Adrenalektomie. *Acta Endocrinol.* **4**, 23.
29. FLÜCKIGER, E. Temperature regulation of hypoxic rats. XVth International Congress of Zoology, Sec. VI, paper 27.
30. FLÜCKIGER, E., AND VERZAR, F. (1955). Lack of adaptation to low oxygen pressure in aged animals. *J. Gerontol.* **10**, 306.
31. FLÜCKIGER, E., AND VERZAR, F. (1952). Senkung und Restitution der Körpertemperatur bei niedrigen atmosphärischem Druck und der Einfluss von Thyreoidea, Hypophyse, und Nebennierenriside auf dieselbe. *Helv. Physiol. Pharmacol. Acta* **10**, 349.
32. FLÜCKIGER, E., AND VERZAR, F. (1953). Überdauern der Adaptation an niedrigen atmosphärischen Druck, nachgewiesen an der Wärmeregulation. *Helv. Physiol. Pharmacol. Acta* **11**, 67.

33. Fregly, M. J. (1954). Cross acclimatization between cold and altitude in rats. *Am. J. Physiol.* **176**, 267.
34. Glaser, E. M. (1949). Acclimatization to heat and cold. *J. Physiol.* **110**, 330.
35. Glaser, E. M., and Shephard, R. J. (1961). Simultaneous acclimatization to heat and cold in man. *J. Physiol.* **156**, 8P.
36. Glaser, E. M., and Shephard, R. J. (1963). Simultaneous experimental acclimatization to heat and cold in man. *J. Physiol.* **169**, 592.
37. Grant, W. C., and Root, W. S. The polycythemic response of sympathectomized dogs to discontinuous hypoxia. USAF School of Aviation Medicine, Project No. 21-2301-0005, Report No. 1, May 1953.
38. Hale, H. B. (1963). Thermal spectrum analysis of thyroid-dependent phases of nitrogen and mineral metabolism. *Federation Proc.* **22**, 766.
39. Hale, H. B., Ellis, J. P., Jr., and Williams, E. W. (1964). Human adrenocortical, sympathoadrenal, and metabolic responses to seasonal changes in climate. *Federation Proc.* **23**, 567.
40. Hale, H. B., and Mefferd, R. B., Jr. (1959). Effects of adrenocorticotropin on temperature- and pressure-dependent metabolic functions. *Am. J. Physiol.* **197**, 1291.
41. Hale, H. B., and Mefferd, R. B., Jr. (1958). Metabolic responses to thermal stressors of altitude-acclimated rats. *Am. J. Physiol.* **195**, 739.
42. Hale, H. B., Williams, E. W., and Ellis, J. P., Jr. (1963). Catecholamine excretion during heat deacclimatization. *J. Appl. Physiol.* **18**, 1206.
43. Hashimoto, K. (1963). The seasonal variation in thyroid function. *Folia Endocrinol. Japon.* **39**, 421.
44. Hiestand, W. A., Stemler, F. W., and Jasper, R. L. (1955). Increased anoxic resistance resulting from short period heat adaptation. *Proc. Soc. Exptl. Biol. Med.* **88**, 94.
45. Higgins, E. A., Iampietro, P. F., Adams, T., and Holmes, D. D. (1964). Effects of a tranquilizer on body temperature. *Proc. Soc. Exptl. Biol. Med.* **115**, 1017.
46. Hudock, A. E., and Berry, L. J. Comparison of glycogenesis and glyconeogenesis in altitude-exposed and normal mice. USAF School of Aerospace Medicine Report 61-44, June 1961.
47. Hudock, A. E., and Berry, L. J. Tissue citric acid and susceptibility to infection in mice during chronic hypoxia and fluoroacetate poisoning. USAF School of Aviation Medicine Report 60-92, December 1960.
48. Hurtado, A., Merino, C., and Dalgado, E. (1945). Influence of anoxemia on the hemopoietic activity. *Arch. Internal Med.* **75**, 284.
49. Johnson, G. E., Sellers, E. A., and Schönbaum, E. (1963). Interrelationship of temperature on actions of drugs. *Federation Proc.* **22**, 745.
50. Jovy, D., Brüner, H., Klein, K. E., and Wegmann, J. M. The problem of cross-adaptation in man. Presented at 13th General Assembly of AGARD, Athens, July 1963.
51. Joy, R. J. T. (1963). Responses of cold-acclimatized men to infused norepinephrine. *J. Appl. Physiol.* **18**, 1209.
52. Kamal, T. H., Johnson, H. D., and Ragsdale, A. C. Environmental physiology and shelter engineering. LVIII. Metabolic reactions during thermal stress (35° to 95°F.) in dairy animals acclimated at 50° and 80°F. *Univ. of Missouri Research Bull.* **785**, January 1962.
53. Lamb, L. E. (1965). Hypoxia—an antideconditioning factor for manned space flight. *Aerospace Med.* **36**, 97.
54. LeBlanc, J., and Rosenberg, F. (1957). Hypothermic effect of chlorpromazine histamine and serotonin, and acclimatization to cold. *Proc. Soc. Exptl. Biol. Med.* **96**, 482.
55. Lewitus, Z., Hasenfratz, J., Toor, M., Massry, S., and Rabinowitch, E. (1964). Uptake studies under hot conditions. *J. Clin. Endocrinol.* **24**, 1084.
56. Lind, A. R., and Bass, D. E. (1963). Optimal exposure time for development of acclimatization to heat. *Federation Proc.* **22**, 704.
57. Mefferd, R. B., Jr., and Hale, H. B. (1958). Effects of abrupt temperature changes

on excretion characteristics of rats acclimated to cold, neutral, or hot environments. *Am. J. Physiol.* **195**, 726.
58. MEFFERD, R. B., JR., AND HALE, H. B. (1958). Effects of thermal conditioning on metabolic responses of rats to altitude. *Am. J. Physiol.* **195**, 735.
59. MEFFERD, R. B., JR., LaBROSSE, E. H., GAWIENOWSKI, A. M., AND WILLIAMS, R. J. (1958). Influence of chlorpromazine on certain biochemical variables of chronic male schizophrenics. *J. Nerv. Mental Dis.* **127**, 167.
60. MEYLER, L., AND PECK, H. M. (1962). "Drug Induced Diseases." Assen, Netherlands: Thomas, 1962.
61. MIYA, F., MARCUS, S., AND PHELPS, L. J. Effect of low ambient temperatures on specific and nonspecific resistance. *In Proc. Symp. Arctic Biol. Med.* III. Influence of cold on host-parasite interactions. (E. G. Viereck, ed.) Arctic Aeromedical Laboratory, 1963.
62. MONCLOA, F., GOMEZ, M., AND HURTADO, A. (1965). Plasma catecholamines at high altitudes. *J. Appl. Physiol.* **20**, 1329.
63. OKAMOTO, M., KOHZUMA, K., AND HORIUCHI, Y. (1964). Seasonal variation of cortisol metabolites in normal man. *J. Clin. Endocrinol.* **24**, 470.
64. PICÓN-REÁTEGUI, E. Studies on the metabolism of carbohydrates at sea level and at high altitudes. USAF School of Aviation Medicine Report 56-107, November 1956.
65. REEVES, J. L. (1961). Influence of intermittent exposure to simulated altitude on plasma and tissue electrolytes in rats. USAF School of Aviation Medicine Report 61-37.
66. REYNAFARJE, B. Myoglobin content and enzymatic activity of human skeletal muscle—their relation with the process of adaptation to high altitude. USAF School of Aerospace Medicine Technical Documentary Report No. 62-89, November 1962.
67. REYNAFARJE, B. Pyridine nucleotide oxidases and transhydrogenase in acclimatization to high altitude. USAF School of Aerospace Medicine Technical Documentary Report No. 62-88, November 1962.
68. SARGENT, F., II, AND JOHNSON, R. E. The physiological basis for various constituents in survival rations. Part IV. An integrative study of the all-purpose survival ration for temperate, cold, and hot weather. Wright Air Development Center Technical Report 53-484, Part IV, December 1957.
69. SARGENT, F., II, SARGENT, V. W., AND JOHNSON, R. E. The physiological basis for various constituents in survival rations. Part III. The efficiency of young men under conditions of moist heat. Wright Air Development Center Technical Report 53–484, Part III, Vol. II, April 1958.
70. SARGENT, F., II, SARGENT, V. W., JOHNSON, R. E., AND STOLPE, S. G. The physiological basis of various constituents in survival rations. Part I. The efficiency of young men under temperate conditions. Wright Air Development Center Technical Report 53-484, Part 1, June 1954.
71. SARGENT, F., II, SARGENT, V. W., JOHNSON, R. E., AND STOLPE, S. G. The physiological basis for various constituents in survival rations. Part II. The efficiency of young men under conditions of moderate cold. Wright Air Development Center Technical Report 53-484, Part II, Vol. I, May 1955.
72. SARGENT, F., II, SARGENT, V. W., JOHNSON, R. E., AND STOLPE, S. G. The physiological basis for various constituents in survival rations. Part II. The efficiency of young men under conditions of moderate cold. Wright Air Development Center Technical Report 53-484, Part II, Vol. II, May 1955.
73. SELYE, H. (1950). "The Physiology and Pathology of Exposure to Stress. Acta, Montreal.
74. STEIN, H. J., ELIOT, J. W., AND BADER, R. A. (1949). Physiological reactions to cold and their effects on the retention of acclimatization to heat. *J. Appl. Physiol.* **1**, 575.
75. STICKNEY, J. C., AND VAN LIERE, E. J. (1953). Acclimatization to low oxygen tension. *Physiol. Rev.* **33**, 13.
76. TAPPAN, D. V., POTTER, V. R., REYNAFARJE, B., AND HURTADO, A. Mechanisms of natural acclimatization tissue enzyme studies and metabolic constituents in altitude adaptation. USAF School of Aviation Medicine Report 55-98, October 1956.

77. TAPPAN, D. V., AND REYNAFARJE, B. Mechanisms of natural acclimatization tissue pigment studies in altitude adaptation. USAF School of Aviation Medicine Report 56-97, October 1956.
78. TAYLOR, R. E., AND FREGLY, M. J. (1962). Effect of reserpine on body temperature regulation of the rat. *J. Pharmacol. Exptl. Therap.* **138**, 200.
79. THOMPSON, E. M., AND KIGHT, M. A. (1963). Effect of high environmental temperature on basal metabolism and concentration of serum protein-bound iodine and total cholesterol. *Am. J. Clin. Nutr.* **13**, 219.
80. TRAPANI, I. L. Environmental extremes and endocrine relationships in antibody formation. *In Proc. Symp. Arctic Biol. Med.* III. Influence of cold on host-parasite interactions. (E. G. Viereck, ed.) Arctic Aeromedical Laboratory, 1963.
81. TRUMBULL, R. Environment modification for human performance. Office of Naval Research Report ACR-109, July 1965.
82. TURNBULL, T. A. Man-induced hyperpyrexia. (1963). *J. Roy. Nav. Med. Serv.* **49**, 169.
83. VAN LIERE, E. J. (1964). Resistance to hypoxia. *Arch. Internal Med.* **113**, 418.
84. WATANABE, G. I. (1964). Seasonal variation of adrenal cortical activity. *Arch. Environ. Health* **9**, 192.
85. WATANABE, G., UEMATSU, M., AND HORII, K. (1963). Diphasic seasonal variation of the serum protein-bound iodine level. *J. Clin. Endocrinol. Metab.* **23**, 383.
86. YOSHIMURA, H., USAMI, S., MORISHIMA, M., KUWATA, T., OTSUKI, Y., AND TOYOKI, M. (1962). Effect of climatic changes on the secretion of adrenocortical hormone. *Kyoto Clin. Endocrinol.* **10**, 260.

# Comments on Cross-Adaptation[1]

MELVIN J. FREGLY

*Department of Physiology, University of Florida,
College of Medicine, Gainesville, Florida*

"Adaptations are modifications of organisms that occur in the presence of particular environments or circumstances. Physiological adaptations appear within a single individual and constitute changes in its functions" (1).

Adaptations may also be viewed in the physical sense by the realization that abnormal environments or circumstances, either external or internal, impose stresses on animals, and the measurable responses of the animals to these stresses are strains. If it were possible to measure quantitatively all the strains responding to a given stress, we could devise a type of Young's Modulus of Elasticity (stress/strain). The reciprocal of this (strain/stress) would represent the adaptive compliance of animals and might be called an index of adaptation. This index could be of value both in determining the degree of adaptation achieved by an animal after a given period of exposure to the stress and in determining when adaptation is complete. It might also be useful in comparing the degree of adaptation achieved by different species under similar conditions. As to the evaluation of the stress factor in our simple equation, we must depend upon physical instruments, such as the thermometer in the case of exposure to cold. This aspect presents little difficulty. The greatest difficulty in attempting to calculate this index comes in the quantification of the strain factor. For example, many responses or strains have been recorded for various animals subjected to cold. Although many "criteria" for cold adaptation have been added through the years, few attempts have been made to evaluate and correlate these individual strains with the process as a whole, so our understanding of which is primary and which is secondary among them is incomplete. To put the question another way: What is regulated or maintained constant and what is controlled to achieve this constancy during adaptations of all kinds? Are there primary strains to which others may be traced? Will the sum of these primary strains be equal to the strain factor we seek for our index? We have no quantitative answers to these questions at present. In attempting to answer them, it must be kept in mind that the multiplicity of strains for a single stress means that adaptation is not a single process, but a syndrome (1). A suggested beginning approach is to study the kinetics of development and loss of each strain during adaptation and deadaptation to graded intensities of a given stress.

## CLASSIFICATION OF ADAPTATIONS

Heilbrunn (7) has classified adaptations into two types: physical and chemical. Physical adaptation includes adaptation to extremes of temperature, adaptation to changes of pressure (e.g., in fishes), adaptation to solar and ultraviolet

[1] Supported by NASA Institutional Grant NSG- 542- Project 41.

radiation, and adaptation to acceleration, weightlessness, etc. Chemical adaptation includes adaptation to changes in ambient oxygen tension (e.g., exposure to altitude or to oxygen at high pressure), osmotic adaptation, and adaptation to injurious or foreign substances (bacterial toxins, drugs, etc.). One might also consider psychologic or behavioral adaptations in this classification, for they would include aspects of both physical and chemical adaptation. Hence, adaptations are possible at all levels of organization from enzymes to behavior. An overriding influence on each of the types of adaptation mentioned above is a genetic factor which can modify the expected responses or strains accompanying adaptation.

## CROSS-ADAPTATION

Adaptation to a single stressor—e.g., cold air or hypoxia—has generally been studied separately. The possibility that adaptation to a single stressor might set into motion a general, somewhat stereotyped, pattern of responses stimulated some investigators to test for cross-adaptation in intact animals. Can different stresses arouse the same strains? Do the strains overlap? Do they reinforce each other, or can they interfere with each other?

TABLE 1
Some Modifications in Physiologic and Biochemical Processes Which Have Been Studied in Man and Other Mammals Exposed Chronically to Cold Air

| Physiologic process | Modification |
|---|---|
| Mean body temperature | No change, (rat) |
| Basal heat production | Increase (rat guinea pig) |
| Food intake | Increase (rat) |
| Thyroid secretion | Increase (rat, rabbit, mouse) |
| Thickened fur, thickened epidermis | Increase |
| Adrenal cortical secretion | Increase (rat) |
| Ascorbic acid excretion | Decrease (man, guinea pig) Increase (rat) |
| Liver, kidney, testis ascorbic acid concentration | Increase (rat) |
| 17-Ketosteroid excretion | Increase possible (man) |
| Nicotinamide excretion | Increase possible (man) |
| Pentose excretion | Increase (rat) |
| Hemoconcentration | No change (man, rat) |
| Plasma volume | Decrease possible (man) |
| Cardiac output | Decrease (man) |
| Blood pressure | Increase (rat) |
| Plasma osmolality and chloride concentration | Increase (rat) |
| Promptness of vasoconstriction in appendages | Increase possible (man) |
| Liver, kidney, heart, intestine weight | Increase (rat) |
| Lower lethal temperature | No change (rat) |
| Catecholamine excretion | Increase (rat) |

## METHODS FOR STUDY OF CROSS-ADAPTATION OR CROSS-ACCLIMATION

The studies that have been carried out at the whole-animal level have adapted the animals to a particular stressor and then tested certain physiologic or biochemical responses (including survival) to a second, different stressor (2, 3, 6, 8, 9). At present, there is no rational basis to know whether response to a second stressor will be positive or negative. Some years ago, I chose to study cross-adaptation (or, more properly, cross-acclimation) between cold and hypoxia, as well as between hypoxia and cold, partly because of the similarity of some physiologic and biochemical responses to each (3). Some physiologic and biochemical processes modified by both exposure to cold and exposure to hypoxia are shown in Tables 1 and 2. It should be emphasized that this list is not exhaustive. There were a number of similarities in the responses to these two stressors, but there were also a number of differences. The results of my experiments indicated that rats acclimated to cold (5 C) had a reduction in their ability to withstand hypoxia, and those acclimated to hypoxia were less able to withstand cold (3). Therefore, cross-acclimation between cold and hypoxia was negative, as was cross-acclimation between hypoxia and cold. Sundstroem and Michaels (11), on the other hand, were able to show a number of positive cross-acclimations between certain stressors, including hypoxia, tumor transplants, exercise, morphine administration, X-rays, cold, bacterial toxins, and protozoa. Each of these stressors was tested against all the others (Fig. 1).

I have also demonstrated a positive cross-acclimation between the development of hypertension and exposure to hypoxia. Rats whose kidneys are encapsulated with latex envelopes develop high blood pressure within 6–8 weeks. If the rats

TABLE 2
Some Modifications in Physiologic and Biochemical Processes Which Have Been Studied in Man and Other Mammals Chronically Exposed to Hypoxia

| Physiologic process | Modification |
|---|---|
| Mean body temperature | Decrease (rat, mouse, guinea pig, dog, rabbit) |
| Basal heat production | Decrease possible (rat, mouse) |
| Food intake | Decrease (rat) |
| Thyroid secretion | Decrease possible (rat) |
| Adrenal cortical secretion | Increase (rat, man) |
| 17-Ketosteroid excretion | Decrease (rat, man) |
| Ascorbic acid excretion | Decrease (man) |
| Hemoconcentration | Increase (man, dog, rat) |
| Plasma volume | Decrease possible (dog, man) |
| Blood volume | Increase (man) |
| Plasma osmolality | Increase (rat) |
| Cardiac output | Increase (man) |
| Heart rate | Decrease possible (man) |
| Blood pressure | Decrease (man) |
| Liver, kidney, testis weight | No change in liver, kidney (rat); Decrease testis (rat) |

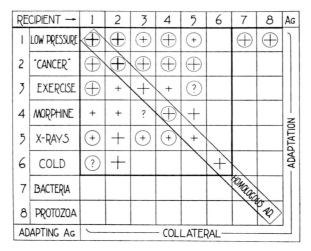

Fig. 1. Adaptation of rats to eight different stresses was studied. After adaptation to a given stress was complete, the responses of rats to the same (homologous) and other (collateral) stresses were tested. Data are from Sundstroem and Michaels (11). Symbols enclosed in circles were obtained by these investigators. The heaviness of the symbols serves as an approximate indicator of significance. Question marks refer to results of doubtful significance. Used by permission of the University of California press.

were exposed to an altitude equivalent to about 12,000 ft immediately after kidney encapsulation, the blood pressure remained nearly normal (4).

Cross-acclimation has also been studied by pharmacologists and biochemists at the enzyme level, although they do not generally view it in this way. It is well known that chronic administration of certain drugs induces an increased synthesis

TABLE 3
Sleeping Times and Hepatic Metabolism of Pentobarbital in Rats Pretreated with $o,p'$-DDD and $o,p'$-DDD + $d,l$-Ethionine for 3 Days

| Treatment | No. rats | Sleeping time, min[a] | Unit activity[b] | Hepatic capacity[c] | Liver:body weight ratio (g/100 g) |
|---|---|---|---|---|---|
| Control | 14 | 86 ± 9 | 146 ± 6 | 6.6 ± 0.4 | 4.6 ± 0.1 |
| $o,p'$-DDD (100 mg/kg) | 6 | 29 ± 3 | 166 ± 8 | 9.4 ± 0.6 | 5.6 ± 0.2 |
| $o,p'$-DDD (300 mg/kg) | 8 | 20 ± 2 | 242 ± 9 | 12.4 ± 0.6 | 5.1 ± 0.1 |
| $d,l$-ethionine (150 mg/kg) | 8 | 137 ± 4 | 108 ± 6 | 4.7 ± 0.5 | 4.3 ± 0.2 |
| $d,l$-ethionine + $o,p'$-DDD | 8 | 78 ± 6 | 131 ± 8 | 5.7 ± 0.3 | 4.4 ± 0.2 |
| | | Probability values | | | |
| Control vs. $o,p'$-DDD | | <.005 | <.005 | <.005 | <.005 |
| 100 mg vs. 300 mg $o,p'$-DDD | | >.25 | <.005 | <.005 | <.05 |
| Control vs. ethionine | | <.005 | <.005 | <.005 | >.25 |
| Control vs. ethionine + $o,p'$-DDD | | >.25 | >.1 | >.1 | >.1 |

[a] Pentobarbital sodium 40 mg/kg ip.
[b] mg × $10^{-3}$/g hour.
[c] mg/kg body wt hr.

of the hepatic enzymes responsible for their metabolism. While the drug in question is metabolized at faster than usual rates after adaptation or induction has occurred, often other, apparently unrelated, compounds are also metabolized at faster rates; i.e., a positive cross-acclimation has occurred. Recently, we have been interested in the physiologic adaptation of rats to chronic administration of the insecticide, o,p'-DDD, an analog of DDT (10). We found that the livers of DDD-treated rats metabolized a standard anesthetic dose of pentobarbital (or hexobarbital) faster than normal, and these rats slept less than half as long as control rats (Table 3). Liver weight was increased by administration of DDD. When ethionine, an inhibitor of protein synthesis, was given with both doses of o,p'-DDD, the decrease in sleeping time was abolished and the mean sleeping

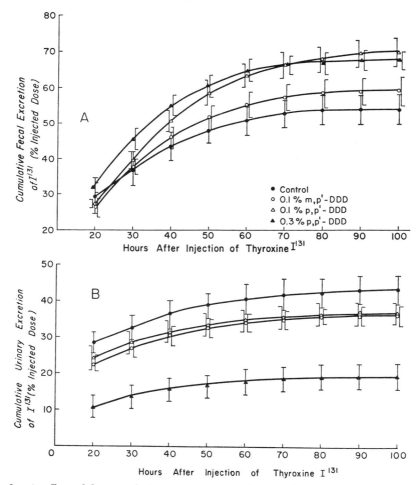

FIG. 2. A, effect of dietary administration of DDD for 6 weeks to female rats on cumulative fecal excretion of $^{131}$I after intraperitoneal administration of thyroxine-$^{131}$I. The mean ±1 standard error is shown for each group of six rats. B, effect of dietary administration of DDD on simultaneous cumulative excretion of $^{131}$I in urine. From Fregly et al. (5); reproduced by permission of the National Research Council of Canada.

time for the experimental animals was not statistically different from the mean control sleeping time. When ethionine alone was given, sleeping times were lengthened such that the mean approached 150% of control time.

At autopsy, we were surprised to observe that rats given $o,p'$-DDD at 0.1% and 0.3% by weight in their laboratory ration for 6 weeks had increases in thyroid weight of 62% and 81% above control level (5). Since the chemical structure of DDD is different from the characteristic antithyroid or goitrogenic compounds, we directed our investigation toward a study of the metabolism of thyroxine by the liver. Administration of thyroxine-$^{131}$I resulted in significantly greater fecal and less urinary excretion of $^{131}$I by all DDD-treated groups than by controls (Fig. 2). In spite of an increase in thyroid weight and an increased metabolism of thyroxine by the liver, there were no changes in the rate of oxygen consumption, growth rate, or other signs of toxicity. DDD-treated rats have an increased thyroxine secretion rate (and larger thyroid gland) because hepatic metabolism of thyroxine increased. The faster turnover of thyroxine is not accompanied by a thyroxine deficiency. In these studies, hepatic enzymes were induced to increase the metabolism of the foreign compound, DDD. Apparently, a nonspecific induction of enzymes concerned with the metabolism of pentobarbital, hexobarbital, and thyroxine also occurred. Thus, predictability of enzymic cross-acclimation between two chemicals is as difficult as predictability of cross-acclimation of the whole animal to two different stressors. Here is the basic problem that must be considered before much further progress can be made in this field. The time is ripe for gathering the large amount of information available regarding the measured responses or strains to given environments or circumstances. Certain gaps in our knowledge will then become obvious and will provide a rational basis for future experiments. For example, questions that may be asked immediately are: "Do enzyme inductions occur in what I have defined as physical and chemical adaptations?" "If so, which enzymes are induced?" "Can this account for many of the changes observed during adaptations of all sorts?" To give a more specific example: "Is the increase in thyroid function observed during exposure of rats to cold air related to an induction of the hepatic enzymes that metabolize thyroxine?"

A possibility exists that a collation and correlation of the presently available responses or strains for given intensities of the various stresses will provide a beginning basis for predicting whether a cross-acclimation will be positive or negative.

## REFERENCES

1. ADOLPH, E. F. (1956). General and specific characteristics of physiological adaptations. *Am. J. Physiol.* **184**, 18–28.
2. FLEISCHNER, J. R., AND SARGENT, F., II. (1959). Effects of heat and cold on the albino rat: crossed resistance or crossed sensitization? *J. Appl. Physiol.* **14**, 789–797.
3. FREGLY, M. J. (1954). Cross-acclimatization between cold and altitude in rats. *Am. J. Physiol.* **176**, 267–274.
4. FREGLY, M. J., AND OTIS, A. B. (1962). Effects of chronic exposure to hypoxia on blood pressure and thyroid function of hypertensive rats, pp. 141–152. *In* "The Physiological Effects of High Altitude" (W. H. Weihe, ed.). Macmillan (Pergamon), New York.

5. FREGLY, M. J., WATERS, I. W., AND STRAW, J. A. (1968). Effect of isomers of DDD on thyroid and adrenal function in rats. *Can. J. Physiol. Pharmacol.* **46,** 59–66.
6. HALE, H. B., AND MEFFERD, R. B., JR. (1958). Factorial study of environmentally-induced metabolic changes in rats. *Am. J. Physiol.* **194,** 469–475.
7. HEILBRUNN, L. V. (1952). "An Outline of General Physiology," pp. 546–563. 3rd ed. Saunders, Philadelphia, Pennsylvania.
8. MEFFERD, R. B., JR., AND HALE, H. B. (1958). Effects of altitude, cold and heat on metabolic interrelationships in rats. *Am. J. Physiol.* **193,** 443–448.
9. MEFFERD, R. B., JR., HALE, H. B., AND MARTENS, H. H. (1958). Nitrogen and electrolyte excretion of rats chronically exposed to adverse environments. *Am. J. Physiol.* **192,** 209–218.
10. STRAW, J. A., WATERS, I. W., AND FREGLY, M. J. (1965). Effect of $o,p'$-DDD on hepatic metabolism of pentobarbital in rats. *Proc. Soc. Exptl. Biol. Med.* **118,** 391–394.
11. SUNDSTROEM, E. W., AND MICHAELS, G. (1942). The adrenal cortex in adaptation to altitude, climate and cancer. *Memo. Univ. Calif.* **12,** 1–410.

# Adaptation to Heat and Cold

Cyril H. Wyndham

*Human Sciences Laboratory, Chamber of Mines, Johannesburg, South Africa*

Man has made a successful cultural adaptation to cold in his use of clothing, shelter, and fire. Consequently, it is difficult to find men, even in primitive communities, who are really cold-stressed. The situation with regard to heat is quite different. Only in highly industrialized communities have effective technologic means been developed by building designers and ventilation engineers of creating artificial microclimates around the individual which are within the comfort zone. These technologic advances have their limitations. There are many situations in which men have to work where it is impossible to provide an artificial microclimate, such as outdoors in summer; nor is it economically feasible to do so in some industries, such as mining at great depth. Consequently, many hundreds of thousands of individuals, in various parts of the world, are exposed to a degree of heat stress which places the temperature regulatory mechanisms under some degree of strain.

This paper will therefore deal mainly with man's physiologic and psychologic adaptations to heat. The psychologists are not represented at this meeting, which I think is a pity, because one of the most important lessons my laboratory in South Africa has learned is that one must examine the reaction of the total organism in considering the health hazards of a stressful environment. If you restrict yourself to the physiologic reactions, you get very limited information, indeed. Furthermore, if you fail to take account of man's psychologic reactions, it may prove to be impossible to implement a well-meaning preventive program.

Consideration of man's adaptation to cold will be brief and limited to an examination of the difficulties in sorting out the contribution to man's "total" adaptation of cultural, psychologic, and physiologic factors.

## ADAPTATION TO HEAT

### Physiology

*Individual variability in reactions to heat.* The Human Sciences Laboratory in Johannesburg has a unique position in the study of human physiologic reactions to heat stress, in that, first, it deals with an industry which has large numbers of men who work at a moderate rate of energy expenditure (because there is little mechanization of mining methods) at very high wet-bulb temperatures (the air in the mines is almost saturated with water vapor due to the fact that the broken rock is, by law, watered at frequent intervals to reduce the amount of dust in the atmosphere), and second, because of its good relations with this industry, it has been able to study large samples of men in its climatic rooms, under controlled conditions of heat stress. These circumstances have led the Human Sciences Laboratory to recognize the fact that, in any large body of workmen exposed to heat stress, there are large interindividual variations in their temperature reactions.

These variations in body-temperature reactions have to be taken into account in the control of acclimatization procedures. Also, any attempt to set limits of heat stress for work in hot conditions which do not take account of this phenomenon is fraught with danger, particularly if the conclusions are based upon the reactions of small samples of individuals who are quite unrepresentative of the population of workmen at risk in an industry. I refer to the studies made on the common experimental animal of the human physiologist in the United States—the ubiquitous male medical student.

Taking first the question of the control of acclimatization procedures, it is probably not well known that large numbers of Bantu laborers are working at a moderate rate at relatively high effective temperatures, underground in the gold mines in South Africa. An estimate for 1964 (26) is the following:

| Effective temperature (°C) | Numbers of men |
|---|---|
| Less than 25.5 | 70,993 |
| More than 25.5 | 103,253 |
| More than 27.7 | 58,650 |
| More than 30.0 | 18,631 |
| More than 32.2 | 3329 |
| More than 33.3 | 498 |

These estimates are based upon "average" air conditions over the year in the working places. In summer, when the greatest risk of heat stroke occurs (75% of the deaths are in the summer months), the effective temperatures in the working places may be 1.0° higher than these averages. This means that 103,253 men would be working at 26.5° effective temperature and higher, 58,650 at 28.7° and higher, and so on.

What adds seriously to the heat-stroke hazard is that, due to agreements with the governments of the regions from which the men are recruited, the Bantu laborers serve, on the average, a 1-year contract. The labor force therefore has a turnover of approximately 100% per annum. In effect, then, there are in excess of 100,000 new recruits put to work each year in effective temperatures in excess of 25.5°. In these circumstances, heat stroke is, potentially, a serious hazard in the gold-mining industry in South Africa. The fact that the hazard has been kept within reasonable limits is due to a number of preventive measures which the industry has adopted. The most important of these is the use of an acclimatization procedure, developed by the Human Sciences Laboratory (17), on all Bantu recruits who are sent to work in places with effective temperatures in excess of 25.5°. Each year, some 250,000 Bantu laborers are acclimatized by these procedures. The procedures are supervised by 130 men who were trained in the Laboratory, examined for their competency, and certified as capable. These men are visited three times a year by the Laboratory staff to ensure that they maintain a high standard of performance in the procedures.

One of the most important problems the Human Sciences Laboratory faced in running the acclimatization procedure was to ensure that the recruits worked hard enough to raise their body temperatures (upon which the development of the state of acclimatization depends), but not at such a high level that they would be in danger of developing heat stroke. The problem was made more difficult by the large interindividual variations in body temperature we found (27). This point is illustrated in Fig. 1 which contains the rectal temperatures of 100 unacclimatized Bantu men, measured after 1, 3, and 5 hours, who were carrying out continuous work at 5 cal/min in an effective temperature of 32.2°. This figure shows that after 1 hour of work the mean rectal temperature was 37.7°, the curve of the distribution of rectal temperatures about the mean was normal, and the highest value was 38.6°. After 3 hours of work, the mean had risen to 38.3°, a distinct skewness to the right had developed, and the highest rectal temperature was 39.6°. By the

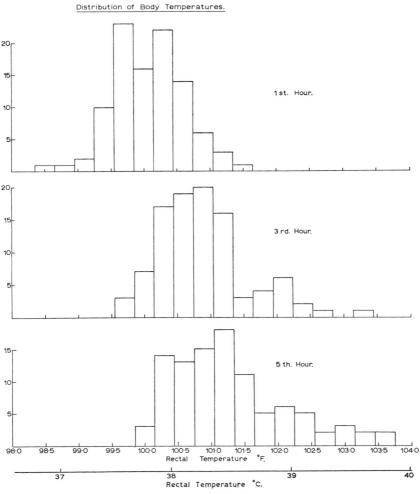

Fig. 1. Distribution of body temperatures.

end of 15 hours of work, the mean rectal temperature had not changed, but the skewness was even more marked and the highest rectal temperature was 39.9°. Thus, although the mean rectal temperature of this sample of men was at a level which would not be considered to be dangerous, some men in the sample, after 3 hours of work, had developed rectal temperatures at which there would be a risk of their developing heat stroke if the rectal temperatures were maintained for some hours.

This example will make it clear that, if we were to safeguard the large numbers of men we acclimatize each year, then some control procedure had to be built into the technique whereby we could recognize those men who are in the skew part of the distribution curve and prevent them from developing very high rectal temperatures. This was done by measuring the oral temperatures of all men undergoing acclimatization before they started work and again after about 3 hours of work on days 1 and 2 of both the cool and the hot stages of the two-stage method. It was found to be necessary to do this on only the first 2 days of each stage, because a study had shown that on day 1 about 50% of new recruits developed an oral temperature of 38.3° and above, but by day 3 this figure had fallen to 3–4%. Men developing an oral temperature of 38.3° (equivalent to a rectal temperature of 38.7°) are immediately stopped from working and are rested for 1 hour. If the oral temperature has not returned to normal after 1 hour of rest, the men are immediately sent to the mine hospital, and the medical officer is asked to examine them thoroughly to exclude any medical cause for the high temperature.

We have been interested in the question of how acclimatization affects the skewness of distribution of body temperature and also the effect of different levels of heat stress on the skewness. Our preliminary evidence shows that in both the acclimatized and unacclimatized men an increase in heat stress causes the mean rectal temperature to increase, the standard deviations to get wider, and the skewness to the right to increase. The means of the unacclimatized men are higher, the standard deviations are greater, and the skewness to the right is more marked. These studies have opened up a new approach to the question of adaptation to heat based on the population reaction, rather than the individual reaction. As we now see it, the main effect of acclimatization is to decrease the danger of heat stroke, not only through its effect in reducing the mean retcal temperature (which is well known), but, even more important, through its effect in reducing the degree of skewness of the distribution of rectal temperatures to the right, and dangerous, side (a phenomenon which this Laboratory has been the first to draw attention to).

These observations on the population reaction to an environmental stress (heat in this case) raise important methodologic questions about our standard approach to the physiologic characterization of health hazards in man's environment. The accent has been on the detailed examination of the biochemical and physiologic reactions of one or two, rarely more than 10, subjects to the environmental stressor. From limited observations on one or two subjects (more often than not, the ubiquitous male medical student), we then have the effrontery and temerity to extrapolate the results to the population at risk, which might be quite different with respect to age, sex, health, physical activities, nutrition, race, etc. I would like to hazard a guess that, if we examined the toxic reactions to drugs, to silica, to radiation, we might also find a skewness in the response of the population to the stressor. One of our aims should be to identify the hyperreactors and study them intensively, because they might contain within their biochemistry and physiology the clues for the environmental stressor's becoming a danger.

The discovery of the skewness in the population body-temperature reaction to the environmental stressor of heat has lead the Human Sciences Laboratory into a number of interesting and important lines of research:

In studies on men, it is obvious that they can only be exposed to a degree of stress which produces sublethal levels of reaction. From such data, the mathematical statistician is asked to estimate the probability of men reaching lethal levels of reaction at various degrees of heat stress; i.e., he has to be able to extrapolate with confidence to a level of body-temperature reaction which is well above the highest level measured. If the distribution of the reactions were normal, this would present a problem; but when the distribution is skew, then the statistical difficulties are compounded. In the case of rectal-temperature reactions, Gumbel's extreme value distribution (developed originally to handle meteorologic data), the log-normal distribution, and the Weibull distribution have all been tried, and each has presented certain methodologic difficulties (7, 8). I have little doubt that the mathematical statistical problems which have been raised by the skewness of distribution of body-temperature reaction to heat stress are not unique to this stressor; if they are looked for, they will probably be found in the population reactions to other stressors, such as drugs.

Another important question raised by the skewness in the distribution of rectal temperatures is the reason for the relative heat intolerance of certain individuals who lie in the tail of the distribution. One wants to know if this is a systematic difference in these individuals from the others. If it is, can the causes be identified? If it is not a systematic reaction of the individual, what are the reasons for the temporary loss of heat tolerance? Some answers are emerging from our studies of the relatively heat-intolerant man. About 10% of the new recruits on the first day of acclimatization are ill (17). Two cases of early lobar pneumonia with rectal temperatures of 104°F were recognized through the control procedures on the first day of acclimatization, and also a case of benign malaria with a rectal temperature of 106°F.

Cumulative dehydration appears to be another common cause of temporary heat intolerance, but we do not yet have enough evidence to be certain of this. There are also certain individuals who are consistently less heat-tolerant than others, and we think that we have a test, the "maximum sweat capacity test," for identifying such individuals. This is done by having men work for periods of 2 hours at a number of different heat-stress conditions of increasing severity. Sweat rates and rectal temperatures are measured and sweat rates are plotted against rectal temperatures. Some individuals show consistently a lower sweat rate with increase in rectal temperature, even when they are highly acclimatized. We want now to look at the number of sweat glands and functional capacity of the glands in such individuals to see whether we can establish a reason for reduced capacity of these individuals to tolerate high heat-stress conditions. This new technique we have developed to measure maximum sweating capacity has led to some interesting results in the control of sweating rates by increase in rectal temperature during acclimatization.

It is not only in the control and the understanding of the effects of acclimatization that these new concepts are important. It is absolutely essential to take account of the skewness in the distribution of the population reaction of body temperature in setting limits of environmental heat stress for men carrying out physical work in hot industries. Failure to do so can lead to misleading and even dangerous conclusions. For example, we (16) and Lind (4) put forward essentially the same physiologic criteria for setting the limits of heat stress for work in industry. These criteria were based upon Neilson's classic observations in 1938 (10) that when a man works at a constant rate his rectal temperature rises from the resting value and reaches a new steady level in about 40–60 minutes. It remains at the new level as long as the man continues to work at that rate. The new level of rectal temperature is directly proportional to the rate of metabolism over a wide range of air temperatures. However, we showed in 1953 that this is true only up to certain critical levels of heat stress. Above these critical levels of heat stress, the body temperature rises to new, higher levels. We also noticed that, the higher the rate of metabolism, the lower the critical level of heat stress.

Based upon 13 highly acclimatized Bantu laborers who worked at three different rates of metabolism for 4 hours, the Human Sciences Laboratory demonstrated that the "average" highly acclimatized Bantu laborer could maintain a steady level of rectal temperature (which was different for the different work rates) up to effective temperatures of 31.7° for light work, up to 30.0° for moderate work, and up to 28.3° for hard work. Based upon three workmen working for 1 hour, Lind stated that the "average" British mine worker could maintain a steady level of rectal temperature up to effective temperatures of 30.0° for light work, up to 27.9° for moderate work, and up to 27.0° for hard work. The lower levels of Lind's subjects are probably due to their poorer state of acclimatization.

On the basis of the "average" rectal-temperature response of these 13 highly acclimatized Bantu laborers, we proposed to the gold-mining industry in South Africa in 1953 (26) that acclimatized Bantu laborers should be able to work, at a moderate rate, up to effective temperatures of about 86°F (30.0°C) without danger of heat stroke. It was soon shown that this limit was too optimistic. In the 5-year period, 1956–1961, there were 15 fatal and 53 nonfatal cases of heat stroke in effective temperatures below 86°F. The failure of the limits based upon the "average" rectal temperature reaction of 13 highly acclimatized Bantu can be attributed to the following reasons:

(1) The subjects of the experiments were not representative of the population at risk in the mines with respect to health, physical fitness, nutrition, state of acclimatization, and possibly age.

(2) The numbers of men studied were inadequate even to give a good estimate of the probable mean temperature reaction of the population at risk. The numbers were quite hopeless for any estimate of the variance in rectal temperature from which tolerance limits could have been calculated, based upon normal distribution

theory (these would also have been incorrect because of the skewnesses of the distribution).

(3) Undoubtedly, the most impotant reason for the failure of the limits was the fact that at that time the Human Sciences Laboratory did not know that the distributions of rectal temperatures become skew after a few hours of work in hot conditions, and that any estimate of the probability of men reaching dangerous levels of rectal temperature which does not take this skewness into account will underestimate the risks of heat stroke seriously.

From our attempts in the Human Sciences Laboratory to characterize the health hazards due to environmental heat stress, we would urge strongly that attention be given to the study of large samples of men under the influence of different stressors, and that, if the distributions of the measured physiologic reactions are skew, as is suggested by these studies, then attention should be given to the development of adequate statistical theory for the prediction of stressor levels at which lethal effects can be expected and of the probabilities of these effects. In the elucidation of the mechanisms of the toxic effects, it may be rewarding to study individuals who are relatively less tolerant to the stressor than just to take subjects at random. Finally, in trying to give limits of the stressor which can be tolerated by the population at risk without danger, it is imperative to take account of the distribution of physiologic or biochemical reactions, so that the limits can be based upon reliable statistical estimates of the probabilities of lethal effects in the population at risk.

*Ethnic differences in adaptation to heat.* A great deal is known about the mechanisms of acclimatization, such as the sweat glands and the circulatory system. By contrast, very little is known about the variations in heat reactions of different ethnic groups living in similar climates in various parts of the world, or of the same ethnic group living in different climates. The motivation for such studies is obvious. The tropical and desert regions of the world could, if developed to their full industrial and agricultural potential, maintain much bigger populations than at present. A population explosion is predicted for the end of the present century. A high priority should therefore be given to the development of industry and agriculture in tropical and desert regions. In fact, this could be one of the biggest challenges to the western world in the latter part of this century. A knowledge of the rates of physical effort that indigenous peoples in these regions and "expatriate" Caucasians are capable of is essential in the planning of these developments.

In order to provide the information required, studies are needed of the heat reactions to a constant work rate in an agreed environmental heat stress of peoples in different parts of the tropical and subtropical regions, in the desert and semi-desert areas, and in temperate climates. By this means, a comprehensive picture could be built up of the influence of race and of local climate on the degree of adaptation to heat. Ten years ago, virtually no information existed, except for the excellent studies of Robinson *et al.* (14) in the 1940's on the heat reactions of Negro and white sharecroppers in southern United States.

Dealing as it does with workmen of different ethnic origin and also with Bantu who are recruited from all over the southern parts of Africa, the Human Sciences

Laboratory has, naturally, been interested in the extent of adaptation to heat of men of different ethnic origin.

It was therefore decided in the Human Sciences Laboratory toward the end of the 1950's to embark upon a series of studies of the heat reactions of different ethnic groups. The first problem the Laboratory had to solve was the combination of work rate and environmental heat stress it should use in these tests. It was decided that the work rate should be representative of the rate of men engaged in manual tasks and that the environmental heat stress should be equivalent to that which men would experience when working in direct solar radiation in tropical or desert regions. To satisfy the first criterion, a work rate of 5 cal/min (equivalent to an oxygen consumption of 1.0 liters/min) was used, because this is the rate which is generally regarded as reasonable in Europe for men engaged in physical work in heavy industry. The climate chosen was one which men working in direct solar radiation in the hot, humid tropics or the deserts might be expected to experience. An effective temperature of 32.2° fulfilled this criterion.

The Human Sciences Laboratory is strongly of the view that studies of this nature must be carried out where the people live. The reason for this desideratum is that it is well known that the state of acclimatization to heat is very labile (5). The level of physical activity, the dietary pattern, and the sleep rhythms of the subjects, especially primitive men, would undoubtedly be interfered with if the subjects were transferred from their homes to a town. These changes are certain to interfere with the men's physiologic responses to heat. To meet this desideratum, a portable climatic chamber was required which could be transported easily and which could be set up and put into working order within a few hours. As funds for these studies were very limited, it had to be inexpensive, too. These conditions were met by constructing a climatic tent with a glass-fiber-filled double wall, which operates at almost saturated air conditions (this avoided the need to have expensive and bulky refrigeration equipment). Experience of the climatic tent constructed in this laboratory (9) showed that it was possible to control with ease to within ±0.5° of the design air conditions of 34° D.B., 32.2° W.B., and air velocity of 80 ft/minute.

With the equipment described and the experimental conditions decided upon, measurements have been made of the rectal temperatures, heart rates, and sweat rates of samples of 20 young adult males of the following groups:

(1) Caucasians in the unacclimatized state in Johannesburg, in winter, and highly acclimatized in the Laboratory; Caucasians living in the hot, humid tropics at Weipa on the Cape York peninsula of Australia and in the Sahara desert at Hassi-Messaoud (Figs. 2 and 3).

(2) Bantu in the unacclimatized state in Johannesburg, in winter, and in the highly acclimatized state (Fig. 2); Bantu from the following areas of southern Africa: Barotseland, Bechuanaland, Nyasaland, Mozambique, Transkei, Zululand, and Angola.

(3) Bushmen living a nomadic, hunting life in the Kalahari desert and "river" Bushmen from the Okavango swamps (Fig. 4).

(4) Aborigines living in the Cape York peninsula of Australia (Fig. 5).

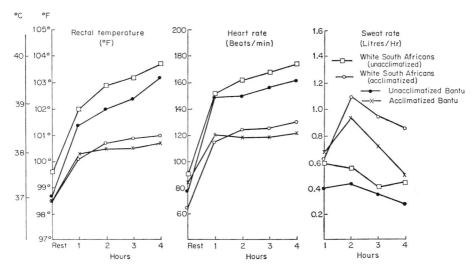

FIG. 2. Caucasians in the unacclimatized and acclimatized state.

(5) Arabs of the Chaamba tribe living at Hassi-Messaoud in the Sahara desert (Fig. 6).

Certain general conclusions can be drawn from the studies on these different ethnic groups (18–24). One is that the climates the different ethnic groups live in play a much more important role in their reactions to the standard heat stress than do differences in body size. This suggests that, if the Bergman rule applies to man, then its effect is very small. Another is that life in a hot desert or in a hot,

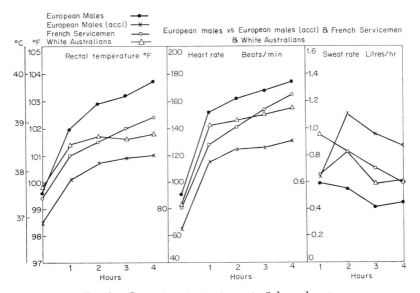

FIG. 3. Caucasians in tropics or in Sahara desert.

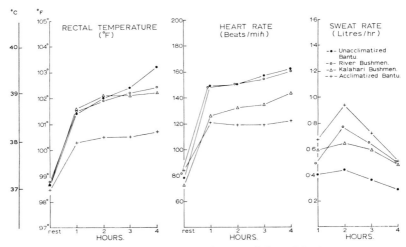

Fig. 4. Bushmen hunters in desert and "river" bushmen.

humid tropic confers about 50% acclimatization (judged on the comparison of rectal temperature, heart rate, and sweat rate of the peoples living in these areas with those of unacclimatized and highly acclimatized Caucasians and Bantu). Other conclusions are that the levels of physical activity of the peoples play a small but significant role in their reactions to the standard heat stress; that the differences in reactions between summer and winter are also significant, but small; and that in the Bantu there were no measurable differences in heat reaction between different tribes living in mild subtropical and temperate regions.

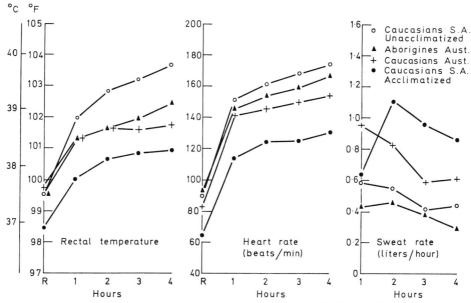

Fig. 5. Aborigines living in Cape York peninsula of Australia.

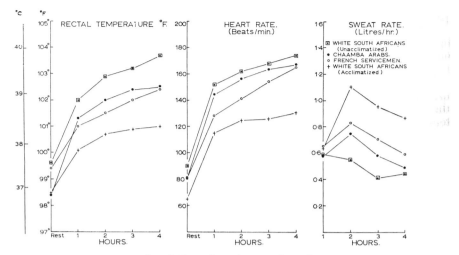

FIG. 6. Arabs of Chaamba tribe in Sahara desert.

These conclusions have far-reaching implications for our understanding of the effects of climate and of race on the heat adaptations of peoples of different ethnic stock. They need to be validated by further studies which are more carefully designed. For example, in the present series, only in the Caucasians were four different samples studied: unacclimatized, desert, hot humid tropics, and Laboratory-acclimatized. If possible, samples of other ethnic groups over a similar range should also be studied. Furthermore, the samples studied were not satisfactory—from the point of view either of their sizes or of being truly representative of the peoples living in the region studied. No doubt, there would be formidable difficulties in the way of keeping to an ideal experimental design. For example, it is unlikely that it would be possible to find "matched" samples of Arabs, aborigines, Bantu, and Caucasians living in the hot, humid tropics, the desert, and temperate regions. However, planning on a broad international scale might help in locating suitable samples of the different populations for these studies.

Consideration should also be given to conducting these studies over a range of heat-stress conditions. An experimental design of this nature should satisfy two further purposes. One is, by taking an arbitrary limit of body temperature, say 40°, to determine the limits of heat stress at which men could work at a moderate rate of physical effort, say 5 cal/min. Such information would be useful in comparing the heat reactions of different ethnic groups and in demonstrating to the governmental agencies, which give the funds for research, that these studies have a practical aim—indicating the limits of heat stress for different ethnic groups carrying out a moderate rate of physical work in different climates. The other purpose would be to obtain information on one of the important temperature-regulatory mechanisms, sweat rate/core temperature. If the sweat rates at four different heat-stress conditions are plotted against rectal temperatures, some idea of the response characteristic of this temperature-regulatory mechanism could be obtained. These curves give the thermoneutral temperature; the slope of the curve, i.e., the sensitivity of the response; and the maximum value of sweat rate, i.e., its capacity.

Recent studies in the Human Sciences Laboratory (Fig. 7) have shown that useful comparisons can be made of men in different states of acclimatization with the following four heat-stress conditions: 22.2, 27.7, 30, and 32.2° effective temperatures.

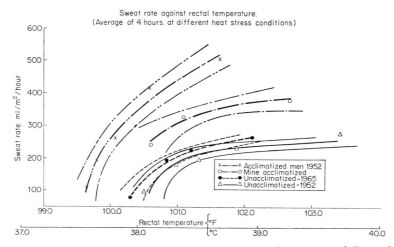

Fig. 7. Sweat rate against rectal temperature (average of 4 hours at different heat stress conditions).

## Psychology

*Behavioral and EEG changes.* The main psychologic feature of man's first exposure to heat, or of the acclimatized man's reaction to extreme heat, is a behavioral change. This takes the form of aggression, hysteria, or apathy. Aggression may be so marked that the subject may assault the scientific observer and it may be impossible to keep the subject in the climatic room. The hysterical subject may moan and weep, thrash about on the floor, foam at the mouth, and hyperventilate. The apathetic patient withdraws into himself and refuses to take part in any group activity or to communicate with the observers. In the female, a loss of inhibitions is observed in addition to the above changes. Normally shy and reserved girls toss aside their clothing in front of male scientific observers, because they find the clothing, scanty as it is, an intolerable burden. Some of the manifestations are well illustrated in the case history of an experienced psychologist on exposure to severe heat:

> In the first half-hour I worked easily and thought that the severity of the stress I had been told I would experience had been exaggerated to me. Discussion with other members of the experimental group and the scientific observer was free, with lighthearted banter. Towards the end of the hour the picture began to change. I felt hot, my face was flushed, and I became aware of my breathing although I was not panting. Conversation dwindled to requests for water, the time, etc. I was pre-occupied with the passage of time, hoping it would pass more quickly than it appeared to do. This feeling was apparently general

among the subjects. My interests narrowed down to task in hand, the passage of time and the thoughts of some escape which would allow me to stop without too great a loss of "face." My body felt remarkably weak and tired (I had once felt like this in a severe febrile illness), but there was no specific anatomical region where I could locate this feeling. My heart beat was now perceptible and when I spoke I had to take a breath between words which I realized was paradoxical because my respiration was not increased but I felt that I had to struggle to get enough air into my lungs and the inhaled air felt hot and oppressive. Talking was a distinct effort. The control of my social inhibitions began to slip about this time. An attempt on the part of the observer to start a conversation was resented and I used violent language to him which I normally would not do. This was apparently general because other members of the team behaved aggressively and childishly when instructions were given. However, when presented with a psychological test it was a relief to focus one's attention away from one's pre-occupation with one's discomfort. Little things caused irritation out of all proportion such as the paper upon which I wrote the psychologist's test becoming wet and difficult to write upon. I realized that I was hyper-reacting to trivial irritations but could not control myself. About this time the discomfort became unbearable. My entire world narrowed down to the single activity of continuing with the task. Doing so seemed to take all my mental effort. Thought stood still. I longed for the brief break for weighing but the effort of starting to work again after the break was very great. I now wanted to hold on only a little longer so as not to show up too badly compared with the other subjects. It was an immense relief when the first of the subjects stopped. I knew that I could give up without too great a loss of prestige.

An endeavor has been made by the Human Sciences Laboratory to see whether any psychophysiologic reasons could be established for the behavioral changes. Electroencephalograms were recorded on 55 subjects before they were subjected to severe heat stress (11). EEG recordings were repeated on 41 subjects immediately after they were withdrawn from the climatic chambers either because they had reached a rectal temperature of 40° or because they were so aggressive that they could not be kept in the climatic chamber. A further EEG recording was taken on 34 men after they had cooled down and their rectal temperatures were near normal. Eleven percent of the control EEG's showed some abnormality, but this is within normal limits.

Of the men who were withdrawn with rectal temperatures of 40° or above, 50% showed apathy, aggression, or hysteria. Three others were withdrawn because they were so aggressive that they could not be kept in the climatic room. One of these men had an EEG which is consistent with epilepsy, although the man had no history of epileptic seizures. All of the cases that were aggressive showed a slowing of the alpha waves during exposure to heat which disappeared when the men had cooled down. Apart from these abnormalities, there were no EEG changes in the men when they were hyperpyrexial or later when they were cooled. It was not possible for a study of the EEG's of the men to establish any psychophysiologic cause for the changes in behavior of the men when exposed to heat stress.

There are a number of other issues that need to be followed up. One is whether heat stress merely brings out an underlying pattern of psychopathology in the individual. For example, does the person with manic-depressive tendencies manifest aggressive behavior on exposure to severe heat stress? Does the individual with schizophrenic tendencies show apathy? Does the individual with psychoneurosis show hysteria? Another is whether the particular behavioral change observed in the individual is specific to the stressor—i.e., heat in this case—or whether similar behavioral changes would be manifest under other types of stress, such as sleep deprivation or anoxia. Yet another is that there appear to be differences between ethnic groups in the predominant features of the behavioral changes. The Bantu under severe heat stress appear to have a higher proportion of men with aggressive tendencies than the Caucasian, although the Human Sciences Laboratory does not yet have sufficient evidence on this point to be certain of it. If this proves to be the case, then it would agree with the conclusions of an experienced psychologist, de Ridder, who has considerable experience of the Bantu and who states: "A third function which came out was a strong aggressive tendency in the Bantu. The interesting feature is not that the Bantu is basically more aggressive than the White man but that this aggression changes from a relative calm to a crazy frenzy of killing in a flicker of an eyelid. His moderation and control elements are far less than in the white man" (2). This aspect of the psychologic reactions to heat of different ethnic groups has not been explored to any extent, as far as I am aware.

There are also important changes in the behavior of the group under heat stress which have been remarked upon by psychologists but have not been explored further. Subjects of an experiment under severe heat stress tend to form an in-group which resents the out-group, comprising the scientific observers. The in-group is hostile to the out-group and, unless there is a sympathetic understanding of the situation by the scientific observers, the stresses set up between the groups can lead to tense situations and might lead to assaults. In one study in which the men were exposed to severe heat for more than 24 hours, the group tried to break out of the climatic room. It is of interest and importance that the men reacted favorably to the voice of one of the senior officials in the mine with whom they were familiar and who spoke to them in their own language. Good leadership, even in these adverse circumstances, restored the morale of the men and they continued with the experiment to its end.

In this regard, one has been impressed with the fact that the recent outbreaks of violence in the slums of the big towns in the United States have generally occurred at times when the heat has been extreme. It may repay someone to examine the correlation between violence and effective temperature, so that civic authorities could be given a guide as to the heat-stress conditions at which great caution is needed to avoid any incident that sparks a riot.

Further research is needed into the behavioral changes of the individual and of the group under severe heat stress, because there are circumstances where exposure to such stresses is unavoidable, such as in military operations in the tropics or the deserts, or in the slums of big cities in summer.

*Chronic exposure to heat.* A number of minor mental changes have been described in men, accustomed to western ways of living and air conditions, when they are transported to tropical regions and remain there for some months. The conditions have been described variously as "tropical fatigue," "tropical neurasthenia," etc.

The clinical manifestations have been well documented (3). True psychotic reactions are rare. The usual disturbances are those which are seen as mental reactions to any stress situation. They are not specific to heat, but also occur when individuals are isolated from familiar surroundings, as in the stress of battle, in situations where the job requirements exceed the individual's abilities, etc. Expatriate white populations in the hot, humid tropics are subject to many unfamiliar sociologic stresses which call for unique powers of adaptation. Those who are unable to adjust are unhappy and frustrated. Such persons often look for some obvious source of annoyance in their surroundings upon which to vent their dissatisfaction. Hot days and nights are an easy and neutral subject to blame. Heat may therefore be substituted for the real cause of discontent in an unhappy individual or community. The part played by heat in the mental disturbances seen in expatriate populations in the hot, humid tropics is in fact very difficult to distinguish from other sociologic factors which themselves, in combination, might provoke similar mental disturbances.

Many of the sociologic factors contributory to this state have been commented upon. A vivid, if gloomy, picture of these stresses was given by Stannus (15) in 1927 of a typical "victim":

> Exiled from home; often separated from his family; generally unable to make ends meet for one reason or another; suffering loneliness and lack of congenial company; envious of others; disappointed over promotion; with ambition thwarted. Living amidst a native population which causes him annoyance at every turn because he has never troubled to understand its language or psychology. From early morn till dewy eve he is in a state of unrest—ants at breakfast, flies at lunch, and termites for dinner. Beset all day by a sodden heat, from which there is no escape, and the unceasing attentions of the voracious insect world, he is driven to bed by his lamp being extinguished by the myriads of insects which fly at night, only to be kept awake by mosquito bites and raucous jungle noises.

Other contributory causes, as commented upon by Ellis (3), are boredom, monotony, isolation, domestic stress, social stresses, and lack of cultural amenities. However, it is probably true to say that, like the impact of all stresses, the seed of discontent will be sown if the soil is weak; a well-adapted person might rise above adversity of this nature. Comment is needed, however, on a clinical observation made repeatedly to me by medical officers with experience of living in tropical communities but, so far as I am aware, not documented by scientific studies. It is that certain personality types do not "thrive" in hot, humid climates, whereas others do; the latter in turn are miserable in the cold and wet of the temperate regions. Psyches which break down in heat are analogous to skins of very fair or red-headed individuals which do not stand up to being continuously moist. Hearsay evidence has it that individuals with psyches which break down in heat are not solely those with "weak personalities"—this suggests that there might be a particular type of personality which cannot stand up to unremitting discomfort due to heat. It is a challenge to the experimental psychologist to identify accurately this type of individual; possibly the study of personality changes during exposure to severe heat might throw light on this question. In the study reported on personality changes of students in heat, an interesting correlation between the assessments of personality in the Pauli test and "breakdown" in heat was revealed. A similar approach might be rewarding in selecting personnel for work or military service in the hot, humid tropics. Considerable cost in transport and frustration to the individual might be saved if a test with a high predictive capacity could be found which would identify those persons with a high probability of manifesting mental disturbances in tropical climates.

Apart from general sociologic factors which contribute to these psychic disturbances and the matter of the individual's threshold, there are two specific factors which have an important bearing. The first is the degree and type of thermal discomfort which leads to these manifestations. Few or no quantitative data exists on this question except in a broad general way. Mental disturbances are rare in the hot, dry climates, but are more common in hot, humid conditions, which suggests that cool nights make bearable the heat of the desert in the day. The intermittent nature in the hot, humid tropics of severe discomfort, which is relentless during both the day and the night in periods of 10 days and is followed by a week of relatively cooler weather during heavy rains, apparently is more disturbing to the psyche. The shorter cycles of discomfort and comfort with a time-period of 24 hours in the desert contrast with the slower cycle of about 10 days in the hot, wet tropics. The second factor which has an immediate effect is skin disease. The effect on the individual's morale of month after month of one boil after another or of a chronic intertrigo that will not heal is quite disproportionate to the extent and severity of the local lesion. It is not difficult to see how the combined and cumulative effects of the two specific factors and some of the background contributory causes can lead to the mental disturbances mentioned and a deterioration of the individual's morale.

It is a mistake to characterize this syndrome as one of fatigue. It is, in fact, a state of mild psychoneurosis with certain specific and certain contributory factors in the etiology, of which climate is a main specific cause. Perhaps one of the most difficult features of the situation is to assess its effect on the individual and on the community and to distinguish the effects of climate, unequivocally, from those due to contributory factors, which can also cause these mental disturbances and lead to a lowering of morale of the community.

One approach which is applicable to the individual is to elicit by interrogation the presence or absence of the various mental disturbances listed at the beginning of this section. Another approach is to examine the attitudes of the community by means of a survey on a random sample of the population. The aim would be to assess the level of morale of the community and express it in quantitative terms by means of the following measurements: a rise in the volume of complaints, an increase in the number of resignations, an increase in the applications for sick leave, an increase in the time taken for standard tasks, an increase in the accident rate, poor quality of workmanship (need to redo work), an increase in the wanton destruction of company property, a higher incidence of psychosomatic ailments and of psychoneurosis, and a falloff in the number and quality of spontaneously organized social events and in attendance at them.

Trends can be followed in these measurements and winter months can be compared with summer. Two of the effects of heat which might be expected to affect these measurements are spells of hot conditions that exceed the comfort limit and an increase in the incidence of illness associated with heat, viz., skin disease and psychoneurosis.

Results of studies of this nature would need to be interpreted with respect to a baseline level of morale of the community. These assessments would not be a simple task. Social relationships in a community are dynamic, and it can be conjectured that in an isolated self-contained community they will be more delicately poised than in or near a town where there is some escape from day-to-day contact with the same individuals. Reliable results would probably require the skilled assistance of a person trained in social psychology, who has been associated with the community for some considerable time. Because of the difficulties in distinguishing between sociologic influences and those due to heat, it is felt that conclusions, given in reports on tropical living, on a "reduction in efficiency," and on "frustration and discontent," should be accepted with some caution where they are based on opinions of persons who are members of the communities involved and, particularly, where the conclusions are unsupported by quantitative data.

*Performance of mental and physical tasks.* Although some studies had been made of the effects of heat stress on the accuracy with which men can carry out tasks, it was in the Applied Psychology Unit of the Medical Research Council at Oxford that careful studies under the direction of Mackworth (6) and then Pepler (12, 13) were carried out on this subject. The main conclusion was that a significant deterioration occurred in all the tasks they studied at effective temperatures of 28–30°. The tasks studied varied from the relatively simple experiment of guiding a pointer, loaded to 50 lb, in a pursuitmeter to the complex experiment of coding messages.

The conclusions recorded by Mackworth and Pepler were never validated by them in the practical work situation, and caution was needed in applying their results to industry. The two reservations one has about the application of the conclusions without proper validation are that the experimental tasks differ from industrial work in that they were all synthetic when seen in the context of machinery used and routines of work in industry, and that the sociologic setting of the subjects in the heat-room studies is quite artificial, compared with the complex sociologic relationships which exist between the workmen and their associates in the factory. The motivation of the men in the experimental heat rooms would therefore be most unlikely to be the same as that of workmen, and recent studies from our Laboratory (25) have shown that the level of a man's productivity is closely related to the level of his motivation for the task in question.

Bearing in mind these criticisms, the Human Sciences Laboratory in 1959 planned a series of studies over a period of 3 years which aimed at providing the gold-mining industry with reliable information on the effects of changes in wet-bulb temperatures and of wind velocity on the productivity of Bantu laborers. The study was restricted to the effects of changes in these two parameters of heat stress, because these are the two environmental heat-stress factors that can most readily be manipulated in the mines by ventilation engineers.

Since shoveling of rock is one of the most widely performed and arduous physical tasks in the mines, it was decided to concentrate the studies on this task.

In order to ensure that the results from the studies would be applicable to practical working situations in the mines, it was decided not to carry out the experiments in a climatic chamber, but to design the investigations in such a way that they met the following two criteria: (1) The experimental situation and tasks had to be as realistic as possible, and (2) the experimental subjects should be reasonably representative of the normal intake of Bantu laborers to the gold-mining industry.

The first criterion was met by arranging for the studies to be carried out in a simulated development end in a mine. An appropriate site was made available by the management of a very hot and deep gold mine (1). The arrangement enabled a gang of two Bantu working under a Bantu supervisor to shovel broken rock into 1-ton mine cars almost continuously for a shift of 5 hours at any selected wet-bulb temperature in the range 28–35.5° and at wind velocities of 100, 400, and 800 ft/minute. The ranges of wet-bulb temperature and wind velocity chosen covered the range of conditions found in hot mines in the gold-mining industry.

To satisfy the second criterion and to avoid any cumulative effect of heat, which may vary with the time of exposure, two Bantu laborers were drawn each day at random from the normal intake of Bantu recruits to the shaft in that mine. The two laborers were taken as soon as they had completed the period of acclimatization to heat by the Chamber of Mines two-stage method; they spent only 1 day in the experiment and then were allocated to their permanent working places. The first stage of acclimatization was carried out at a wet-bulb temperature of 30°, and the second stage, at 32.2°.

As two additional measures to satisfy the second criterion, groups were drawn at random from the different tribes coming to the mine, although each group always consisted of two men of the same tribe to avoid possible conflicts; and observations were repeated on a number of different groups at each temperature condition to enable the variability between groups in production to be assessed statistically. This measure was essential to allow for the calculation of confidence limits about the average levels of production at each temperature condition from which the significance of differences between the various air velocities, etc., could be tested statistically.

The procedure each day was for the scientific observer to set the air conditions as required in the experimental working place; the oral temperatures and heart rates of the two subjects were then recorded and they received instructions from the Bantu supervisor to the effect that they were to carry out the normal rate of shoveling into the 1-ton cars throughout the shift. The subjects filled the cars with rock but did not tram the full cars out of the working place; this was done by a separate group of men. The supervisor was not permitted to shovel rock himself, but could loosen the rock to make the shoveling easier for his group. He was instructed that the group should shovel as much rock as possible out of the working place in the shift.

The scientific observer kept out of sight as much as possible during the shift and intervened only to measure the men's oral temperatures and heart rates after 1, 3, and 5 hours of work or when the supervisor suspected that one of the subjects was suffering from a heat illness and required confirmation of this. The criterion for stopping a subject from working was an oral temperature of 40° or signs of undue strain or impending collapse. These procedures and the presence of a European acclimatization supervisor are normal in the acclimatization center. It was not expected, therefore, that these factors would alter the motivation of the subjects. The scientific observer recorded the temperatures and air velocities in the working places. He also kept a record of each mine car as it was filled and the time to fill the car.

In processing the data for analysis, allowances were made for interruptions in work, i.e., the time taken to change the mine cars and to measure the oral temperatures and heart rates. Thus, the actual time put into the job was calculated. The men's productivity was then expressed in terms of the number of cars filled per hour of active working time. Stoppages by the men due to fatigue were calculated in the actual working time.

### (1) *Effect on Productivity of Increasing Air Temperatures and Air Velocities*

Accurate records were made of each group's productivity on the task of shoveling rock into 1-ton mine cars at the three wind velocities—100, 200, and 800 ft/minute—and, at each wind velocity, at five wet-bulb temperatures—27.7, 28.9, 30.5, 32.2, and 35.5°. In the study of 100 ft/minute, there were six or seven groups, each of two men, at each of the five temperature conditions. At both 400 and 800 ft/minute, five groups were studied at each of the temperature conditions.

The method of analyzing the data was to plot the mean productivity figures in cars/hour (of the five to seven groups) for each of the five temperature conditions on separate graphs for the 100, 400, and 800-ft/minute air-velocity conditions. A logarithmic transformation of the original data was then made in order to stabilize the residual variances. (Specifically, the transformation was $Z = 2 - \log_e Y$, where $Z$ is the air temperature in °F, and $Y$, the number of cars/hour.) An exponential curve of the form, $Z = a + br^x$, was fitted through the transformed variables by the method of least squares. Then 78% confidence limits could be calculated, satisfactorily, to the fitted curves. Finally, the curve and the limits were transformed back to the original scale.

Comparison of the effects of different wind velocities on productivity over the air-temperature range studied is clearer and the conclusions are more general when the curves of productivity against wet-bulb temperature are expressed as a percentage falloff in productivity from the level at 27.7° wet-bulb temperature. A similar method of log transformation was used to stabilize the residual variances and for the calculation of the 78% confidence limits.

The curves of the percentage falloff in productivity at the three wind velocities are shown two at a time in Figs. 8, 9, and 10, together with 78% confidence limits. The interpretation of these confidence limits is that, where they do not

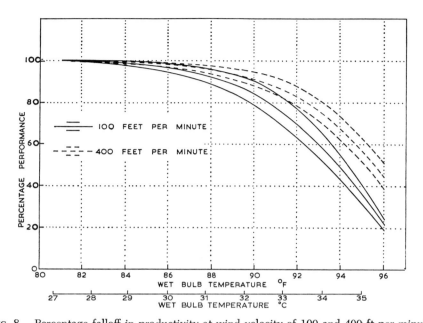

FIG. 8. Percentage falloff in productivity at wind velocity of 100 and 400 ft per minute.

overlap, there is a 95% chance that there is a real difference between the two curves. From these studies, the following conclusions may be drawn:

(1) At a wind velocity of 100 ft/minute a 4% falloff in productivity occurs at 30.0° wet-bulb temperature (taking that at 27.7° as 100%), and the rate of

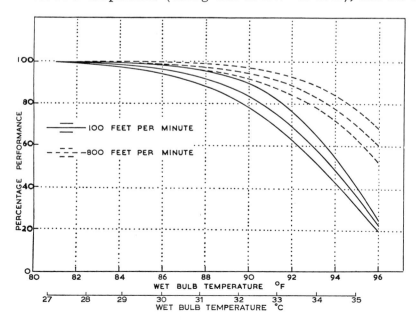

FIG. 9. Percentage falloff in productivity at wind velocity of 100 and 800 ft per minute.

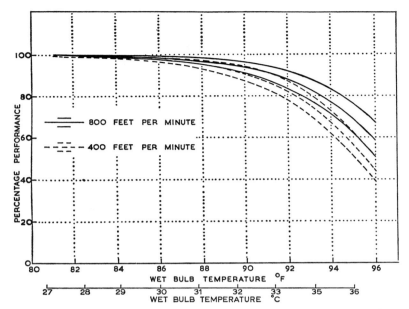

Fig. 10. Percentage falloff in productivity at wind velocity of 400 and 800 ft per minute.

falloff increases with rise in wet-bulb temperature, so that at 35.5° the level of productivity is only 21% of that at 27.7° wet-bulb temperature.

(2) Increasing the wind velocity decreases the rate of falloff in productivity, so that at 35.5° and a wind velocity of 400 ft/minute the level of productivity is 44% of that at 27.7° and at 800 ft/minute it is 59%.

(3) The effect of increasing wind velocity from 100 to 400 ft/minute is significant above 32.8°; from 100 to 800 ft/minute, significant above 31.1°; and from 400 to 800 ft/minute, not significant.

(2) *Effect on Productivity of Quality of Supervision*

The effects on productivity of increasing wet-bulb temperatures from 27.7 to 35.5° at 100 ft/minute wind velocity was studied under three different supervisory conditions. In one, the experiments were made with the subjects under the supervision of a "good" Bantu supervisor. He had been selected on his good production record and reputation as a hard-driving man. In the second, the Bantu supervisor was chosen because he was known to be "inefficient" and had a history of being changed from one job to another. In the third, the groups worked without direct supervision by a supervisor. In the first two supervisory conditions, one group only was studied at each of the five temperature conditions, and in the third supervisory condition, three groups were studied at each of the five temperature conditions. A number of important conclusions can be drawn from these results:

(1) Under relatively cool air conditions (below 30° wet-bulb temperature), the level of productivity of the Bantu laborers engaged in shoveling rock is directly dependent upon the quality of the supervision.

(2) In really hot conditions—i.e., above 33.3° wet-bulb temperature—the effect of heat appears to be so profound that it swamps the effects that differences in the quality of supervision may be expected to have. The significance of this is that good production supervisors would be wasted if they were put in charge of groups working at high wet-bulb temperatures. Because of the dangers of heat illnesses at high wet-bulb temperatures, the ideal supervisor in these circumstances is one who, while maintaining a steady rate of work, is able to detect early signs of distress among the laborers working under his care.

(3) That the relationship between percentage production and wet-bulb temperature is practically independent of the quality of the supervision means that different sets of relationships showing the effects of wind velocity and wet-bulb temperatures do not have to be produced for different qualities in the supervision. Therefore, mine managers can apply these results in a general way without having to take any account of the differences in the quality of supervision in different working places. This finding increases the practical usefulness of the results greatly.

These findings on the influence of the quality of supervision on the effect of heat on productivity are in close accord with one of Mackworth's conclusions (6). He showed that, while highly motivated wireless telegraphers had higher levels of performance in comfortable air conditions than the ordinary operator, the percentage falloff in efficiency with increase in effective temperature was the same in both groups. In Mackworth's case, the motivation came from within the men. In the present study, the differences in motivation were due to the difference in quality of the supervision of the men—i.e., the motivation comes from without the men. In can be concluded from Mackworth's and the Human Sciences Laboratory's studies, therefore, that the effects of heat lead to the same percentage falloff in productivity, irrespective of whether the differences in levels of productivity in comfortable conditions are due to the men's own motivation or to the motivation which comes from the supervisors.

## (3) *Effect on Productivity of State of Acclimatization*

From the studies on the effects of heat on productivity reported above, mine managers are in a position to estimate with some confidence the probable effects of ventilation schemes on the productivity of their manual laborers underground. In general, ventilation engineers can control only two of the heat-stress factors. They can increase the air velocity greatly and drop the wet-bulb temperature a small amount by increasing the volume of air sent underground. This requires expensive fans. They can, alternately, put their money into refrigeration plants and hope to drop the wet-bulb temperature in the working places. Either scheme is expensive in the capital outlay and in the maintenance costs. A third alternative presents itself as a possible means of improving productivity in hot conditions which, if successful, would be of negligible costs to the industry. This is to increase the state of acclimatization to heat by means of a third stage of acclimatization in very hot conditions.

To test this possibility, groups of Bantu laborers were subjected to a third stage of acclimatization in which they worked for an extra week at 33.9° wet-bulb temperature before being subjected to the test. The studies were carried out in exactly the same way as the preceding experiments at the same six wet-bulb temperatures and at a wind velocity of 100 ft/minute. Three different groups were studied at each of the wet-bulb temperatures.

Curves were fitted to the data by the same techniques as described previously and the curve for the percentage falloff in productivity with rise in wet-bulb temperature, with 78% confidence limits, was compared with the curve on the normally acclimatized men. The confidence limits overlap throughout the range of wet-bulb temperatures, and it is quite clear, therefore, that hyperacclimatized men do not have a higher level of productivity at high wet-bulb temperatures.

However, even though there was no significant influence of hyperacclimatization on productivity, these men did show lower oral temperatures. Therefore, we can conclude that this procedure allows men to work at very high wet-bulb temperatures with greater safety than normally acclimatized men.

### (4) *Accuracy of Predicting Falloff in Productivity with Different Heat-Stress Indices*

Mackworth expressed his results on the effects of heat on efficiency of performance in terms of effective temperature, but made no attempt to see whether this index of heat stress takes accurate account of the relative effects of the four heat-stress factors in the environment on human performance. Pepler (13) carried out a short series of experiments on air conditions with the same effective temperature, but with different humidity conditions. His results were, however, inconclusive. It is indeed most surprising that work study and production engineers in mines and other hot industries throughout the world make their estimation of the effects of heat on productivity in terms of the effective-temperature scale and yet not one study, to my knowledge, has been carried out to ascertain whether that scale does, in fact, give correct weighting to the relative effects of the four heat-stress factors on human performance and productivity.

One of the objects in the design of the experiments on the effects of heat on productivity by the Human Sciences Laboratory was to examine the relationship between productivity and both the effective-temperature scale and the wet-Kata cooling powers over the range of air conditions that exist in the mine. For this purpose, the data at the three wind velocities—100, 400, and 800 ft/minute—over the range of wet-bulb temperatures of 27.7–35.5° were used. The first step in this analysis was to fit separate curves, for each of the three wind velocities, to the data of productivity against effective temperature on the one hand, and against wet-Kata cooling powers, on the other hand, in the same way as the curves were fitted for productivity against wet-bulb temperature before. These curves were then used to determine the mean falloff in percentage productivity with increase in effective temperature and decrease in wet-Kata cooling powers for each of the three wind velocities, taking the level of productivity at 22° effective temperature and wet-Kata of 19.8 as 100%.

These results bring out the fact that both the effective temperature and the Kata thermometer take account accurately on the effect of increasing the wind velocity from 400 to 800 ft/minute over the range of 27.7–35.5° wet-bulb temperatures. However, neither takes accurate account of the effect of increasing wind velocity from 100 ft/minute to either 400 or 800 ft/minute in that the 78% confidence intervals do not overlap in the very high-heat stress conditions. The wet-Kata cooling power is less accurate than the effective temperature in this regard, as is shown by the wider range of wet-Kata, than effective temperature, over which the limits do not overlap. Detailed statistical analysis also demonstrated that the significance levels of the differences in the mean falloff in productivity at 100 and 400 ft/minute were higher with the wet-Kata than the effective temperature.

From this analysis, it must be concluded that we do not have at present a heat-stress index which predicts, accurately, the effects of heat on the decrease in productivity. The shortcomings of the effective temperature and the wet-Kata cooling power lie in the fact that they both fail to take account accurately of the very important effect of increasing wind velocity from 100 to 400 ft/minute. It is this range of air movement that is most important to industry and where accuracy in the prediction of the effects of heat is essential.

This examination of the relationship between percentage falloff in productivity and effective temperature allows us to compare our results with those of Mackworth or Pepler in the climatic-room studies, as they expressed their results in relation to effective temperature of the environments. If we take a mean fall in percentage productivity as significant when the 78% limits fell below the 100% productivity level, then a significant falloff is noted at approximately 27.5° effective temperature, and at 30.0° effective temperature the level has fallen to about 10% below the 100% level of 25.4°. Mackworth and Pepler also found that efficiency in the tasks they investigated fell off significantly between 27.7 and 30.0° effective temperature.

## Practical Considerations

Given the situation where large numbers of men are urged on in their physical efforts by their supervisors, as is the case in the mines in South Africa, or by their sergeants, as is the case in the armed forces, the first concerns of the individuals responsible for the men should be: (1) To have available an index of heat stress which is simple to use, is accurate, and gives the correct weighting to the relative physiologic and psychologic effects on the population at risk of the various heat-stress parameters as they occur in the men's environments; and (2) to have statistically reliable estimates of the probabilities that men, in various states of acclimatization, will reach dangerous levels of body temperature at the levels of heat stress which occur in the conditions the men will be exposed to, and estimates of the decrease in efficiency and the fall in productivity in this range of heat stress.

In the view of the Human Sciences Laboratory, there is at present no index of heat stress which can be used to predict, reliably, the physiologic reactions of men engaged in physical work, at different rates, in various conditions of heat stress. It is also true that the physiologic criteria used for judging when men either have reached the limits of their tolerance to heat stress or are in danger of heat stroke are still a matter of contention among physiologists.

The Human Sciences Laboratory is engaged in a large-scale mathematical-statistical exercise in which the data which it has collected since its inception in 1950 on large samples of Bantu, in various states of acclimatization, under different combinations of heat-stress parameters, are being analyzed (1) to determine the relative accuracy of the presently available heat-stress indices and scales for predicting the probabilities that men at different levels of heat stress will reach dangerously high body temperatures and for estimating the fall in productivity; and (2) to give statistically reliable estimates of the probabilities of dangerously high levels of body temperature and of the falloff in productivity at various levels of heat stress.

With the opening up of new mines in hot regions, it was imperative for the gold-mining industry in South Africa to be given urgent guidance on the second of these two questions. In order to give estimates of the risks of heat stroke at the heat-stress levels in the mines, the Human Sciences Laboratory obtained figures on the numbers of Bantu laborers working at certain class intervals of wet-bulb temperature, as they occur in the mines, for each year of the 5-year period 1956–1961 (26). Figures were also obtained for that period of the numbers of fatal and nonfatal cases of heat stroke in the same class intervals of wet-bulb temperature. From these data, it is possible to estimate the incidences of fatal and nonfatal heat stroke for the different class intervals of wet-bulb temperature and to fit curves to the estimated values of incidences of fatal and nonfatal heat strokes against wet-bulb temperature (Table 1).

The wet-bulb temperature class intervals in the table can be converted into class intervals of effective temperature by using the figure for the "average" velocity of air movement in the working places in the mines of approximately 100 ft/minutes.

On this basis, curves have been drawn in Fig. 11 for the probabilities of fatal and nonfatal cases of heat stroke at various effective temperatures in the gold mines in South Africa. From these curves, it will be clear that, below 25.5° effective temperature, the risk of heat-stroke cases is negligible, and that, at 27.7° effective temperature, the risk of fatal heat stroke begins. Up to 32.2° effective temperature, there is a slow, steady rise in the risks of both fatal and nonfatal cases of heat stroke, but above 33.3° effective temperature, the risks increase sharply.

It should be borne in mind that these curves apply only to men in the high state of acclimatization of the Bantu in the gold mines in South Africa and to men carrying out physical work at a moderate rate under direct supervision.

TABLE 1
INCIDENCE OF HEAT STROKE IN PERIOD 1956–1961

| Temperature (°F) | Total no. | No. fatal | Fatal/1000 | No. nonfatal | Nonfatal/1000 |
|---|---|---|---|---|---|
| 80 | 371,318 | 0 | 0.0000 | 3 | 0.0081 |
| 80–83.9 | 177,960 | 0 | 0.0000 | 6 | 0.0337 |
| 84–87.9 | 178,536 | 15 | 0.0840 | 44 | 0.2464 |
| 88–90.9 | 89,113 | 16 | 0.1795 | 62 | 0.6957 |
| 91–92.9 | 17,507 | 10 | 0.5712 | 28 | 1.5994 |

In Fig. 11 are also given curves derived from Figs. 8, 9, and 10, in which productivity, expressed as a fall in percentage production, is plotted against effective temperature for air movements of 100 and 400 ft/minute. This shows that, at 25.5° effective temperature, productivity is at 100%; at 27.7°, it has fallen to 95%; and at 32.2°, it is as low as 70% of the value at 25.5°.

Figure 11 conveys to mine managers both the increase in risks of heat stroke and the probable fall in productivity as the heat stress in the environment increases in the mines. It has led a number of mine managers voluntarily to set limits of environmental heat stress for work in their mines. No doubt, similar curves could be obtained in other hot industries. If heat stroke cases do not occur in the industry, the relationship between some form of heat illness, such

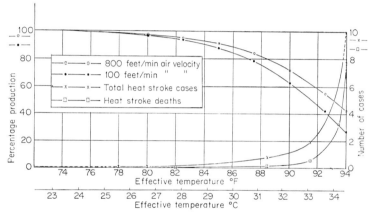

Fig. 11. Incidence of heat stroke deaths and change in productivity with increase in effective temperature.

as heat collapse, and effective temperature could be obtained and would be equally illuminating, as well as being a forceful argument in persuading management to agree to certain limits of heat stress in the environment of the workmen.

## ADAPTATION TO COLD

In man's adaptations to cold, it is much more difficult than it is in his adaptations to heat to sort out the contributions of cultural, psychologic, and physiologic factors to his "total" adaptation to cold. In fact, man has made such a successful cultural adaptation to cold in his wise use of fire, protective clothing, and shelter that it has been very difficult for physiologists to find communities, even among primitive peoples, who are really cold-stressed. For example, the Eskimo lives in the coldest regions on earth, yet according to Milan, the microclimate in his igloo and under his clothing is similar to that of man in temperate regions. It is only under unusual, or emergency, conditions that the Eskimo is cold-stressed.

It will be clear, therefore, that, before the physiologist looks for possible physiologic forms of adaptation in primitive man, a preliminary study should be made of the communities' cultural adaptations and of the resulting microclimates which the individuals are actually exposed to. Physiologists studying primitive communities have been very remiss in this regard. In very few of the reported studies is it possible to gain any precise information on the degree and duration of the cold stresses the different members of the community are exposed to. The importance of this information is illustrated by the cultural adaptation made by the Kalahari bushmen to the cold of winter, when air temperatures may fall below freezing point in the early hours of the mornings and a sharp wind blows across the land. By the wise use of fire, the skin cloak, and a wind break, the bushmen have successfully blunted the sharp edge of the winter's chill. Studies have shown that, by skillful use of fire and the cloak, they achieve an air temperature next to the skin which is in the thermoneutral range. The Kalahari bushmen have no need, therefore, to make any physiologic adaptation to cold, and in fact none is seen in the metabolic and body-temperature reactions over a wide range of temperatures in experiments with 2-hour exposures. Other cultural features which affect man's metabolic and body-temperature reactions are his nutrition and his activity patterns. The Eskimo eats large quantities of protein in his diet, and the influence of this on the raised metabolism during cold exposure has to be borne in mind. Similarly, the state of physical training and the level of physical activity prior to the tests can each produce metabolic reactions during mild cold exposure which could be attributed, incorrectly, to acclimatization to cold.

Another factor making the interpretation of physiologic adaptations to cold difficult is man's psychologic adjustment. Studies of thermal comfort in winter and summer have shown that man's appreciation of temperature sensation is subject to adaptation. For example, Bedford in Britain showed that the "comfort zone" of effective temperature shifts down by about 2–3° in winter compared with summer. It follows that the same air temperature that will cause an individual to be uncomfortably cold and to shiver in summer may not be considered to be cold or produce shivering in winter. Such psychologic adjustments are an important feature of "temperate" man's adaptation to life in the polar regions. This feature also makes it difficult to interpret the metabolic reactions of primitive peoples during sleep in cold conditions. Primitive peoples, such as the Australian aborigines, are accustomed to sleeping under conditions—cold, hard surfaces, etc.—which would cause discomfort in the Caucasian, who would lie awake or sleep only very lightly. Shivering has a conscious element, and therefore it will be suppressed during sleep, as with anesthesia or alcohol. The ability of primitive man to sleep deeply because of his different threshold of discomfort, compared with the Caucasian's, is a psychologic adaptation to cold which makes it difficult to interpret the physiologic reactions observed, such as metabolism and body temperature.

## REFERENCES

1. Cooke, H. M. *et al.* (1961). The effects of heat on work performance. *Bull. Min. Ventil. Soc. S. Africa* **177**.
2. de Ridder, J. S. (1965). The personality of urban Bantu and its tribal roots. *Proc. Min. Med. Officers S. Africa* **65**, 52.

3. ELLIS, F. P. Tropical fatigue. *In* "Symposium on Fatigue" (W. Floyd and A. Welford, eds.). Lewis, London. (1953).
4. LIND, A. R. (1960). The effects of heat on the industrial worker. *Ann. Occupational Hyg.* **2**, 190.
5. MACHLE, W., AND HATCH, T. F. (1947). Physiological responses to heat. *Physiol. Rev.* **27**, 200–227.
6. MACKWORTH, N. H. Researches on the measurement of human performance. Med. Res. Coun. Spec. Rep. No. 268, London. H.M.S.O., 1950.
7. MARITZ, J. S. Estimation of heat stroke risks using extreme value distribution. Chamber of Mines Research Rep. No. 12/65, Johannesburg, 1965.
8. MARITZ, J. S., AND MUNRO, A. M. The use of generalised extreme value distribution in estimating extreme percentiles. Chamber of Mines Research Rep. No. 8/65, Johannesburg, 1965.
9. MINICH, S. G. (1960). An inexpensive portable climatic room. *J. Appl. Physiol.* **15**, 1154.
10. NEILSEN, M. (1938). The regulation der körpertemperatur bei muskelarbeit. *Skand. Arch. Physiol.* **79**, 193.
11. NELSON, G. K. The encephalogram and behaviour under conditions of high temperature and humidity. C.S.I.R. Spec. Rep. No. PERS 81, Johannesburg, 1964.
12. PEPLER, R. D. The effect of climatic factors on the performance of skilled tasks by young men living in the tropics. Med. Res. Counc. London. R.N.P. 53/764, 1953–1954.
13. PEPLER, R. D. (1958). Warmth, glare and background speech—a comparison of the effects on performance. *Ergonomics* **2**, 63.
14. ROBINSON, S., DILL, D. B., WILSON, J. B., AND NEILSEN, M. (1941). Adaptations of white men and Negroes to physical work in humid heat. *Am. J. Trop. Med. Hyg.* **21**, 261.
15. STANNUS, H. (1927). Tropical neurasthenia. *Trans. Roy. Soc. Trop. Med. Hyg.* **70**, 327.
16. WYNDHAM, C. H. *et al.* (1953). Practical aspects of recent physiological studies in Witwatersrand gold mines. *J. Chem. Metall. Min. Soc. S. Africa* **53**, 287.
17. WYNDHAM, C. H. *et al.* (1954). A new method of acclimatisation to heat. *Arbeitsphysiologie* **15**, 373.
18. WYNDHAM, C. H. *et al.* (1964). Heat reactions of Bantu and Caucasians in South Africa. *J. Appl. Physiol.* **19**, 598.
19. WYNDHAM, C. H. *et al.* (1964). Heat reactions of Caucasians in temperate, in hot dry and hot humid climates. *J. Appl. Physiol.* **19**, 607.
20. WYNDHAM, C. H. (1964). Heat reactions of some Bantu tribesmen in southern Africa. *J. Appl. Physiol.* **19**, 881.
21. WYNDHAM, C. H. *et al.* (1964). Physiological reactions to heat of Bushmen and Bantu. *J. Appl. Physiol.* **19**, 885.
22. WYNDHAM, C. H. *et al.* (1964). Reactions to heat of Arabs and Caucasians. *J. Appl. Physiol.* **19**, 1051.
23. WYNDHAM, C. H. *et al.* (1964). Heat reactions of Aborigines and Caucasians. *J. Appl. Physiol.* **19**, 1055.
24. WYNDHAM, C. H. *et al.* (1965). The adaptations of some of the different ethnic groups in southern Africa to heat, cold and exercise. *S. African J. Sci.* **61**, 11.
25. WYNDHAM, C. H. (1965). An operational study of the influence of human and other factors in industrial productivity. *J. Inst. Mech. Eng. S. Africa* **14**, 239.
26. WYNDHAM, C. H. *et al.* (1965). A survey of the causal factors in heat stroke and of their prevention in the gold mining industry. *J. S. African Inst. Min. Metall.* **66**, 125.
27. WYNDHAM, C. H. *et al.* (1960). The temperature responses of men after two methods of acclimatisation to heat. *Arbeitsphysiologie* **18**, 112.

## Commentary

### Harwood S. Belding

*Graduate School of Public Health, University of Pittsburgh, Pittsburgh, Pennsylvania*

Dr. Wyndham is focussing on important aspects of the "economics of ergonomics." Occupational economics involves assurance of well-being of the worker while seeking a financial profit, while ergonomics deals with the arrangement of work to fit capacities of workers.

Although many of Dr. Wyndham's studies relate primarily to the special practical problem of work in hot, very humid conditions, there are important general lessons for us in his approaches. For example, he uses samples of the population at risk as subjects, often observing their responses at the work place, rather than attempting to extrapolate from laboratory findings on the young, fit college students upon whom physiologists usually depend for "normal" baselines. His underground simulation of work at the work site, with laboratory-type controls of wind and temperature is ingenious; I regard his study of effects of these variables on productivity as a classic.

Dr. Wyndham has made progress on two important challenges to thermal physiologists. The first is that of rating particular combinations of hot conditions and work intensity in terms of the relative physiologic strain that is to be anticipated. Dr. Wyndham's success in this derives partly from the fact that he is dealing with a simplified situation, where radiant heat load and humidity are not variables. In the presence of radiation and in lower humidities such rating is more complicated and remains a challenge.

The second challenge is that of predicting, on the basis of a single simple test, the later performance and well-being of individuals being recruited for work in the heat. The problem here is that those of both good and poor potential may show up poorly if tested when unacclimatized, and the capacity for individual improvement with subsequent acclimatization is not obvious. Perhaps the key to success in making a prediction of subsequent successful adaptation simply lies in obtaining evidence of well-developed cardio-respiratory capacity.

# Cardiac Disease in the Context of the Future Environment[1]

Steven M. Horvath

*Institute of Environmental Stress, University of California, Santa Barbara, California*

The large dimensions of the cardiac-disease problems in the United States are clearly indicated in the data obtained by the National Health Survey (5). Approximately one of every four adults has reason to be concerned regarding heart disease. The survey results indicated that more than 14 million adults have some form of heart disease and that 3 million of these have coronary heart disease. Another 13 million adults may have "suspect heart disease." The magnitude of the problem can be stated in another manner: six out of every 100 males and four of every 100 females have coronary heart disease.

According to some geneticists, we cannot anticipate, within the foreseeable future, the elimination through genetic adaptation of individuals with either the tendency to develop cardiac disease or the genetic programming which results in reduced susceptibility to various environmental factors leading to cardiac and vascular disease. We can only anticipate in the next 50–100 years the presence of greater numbers of individuals with proclivity to these disorders. Consequently, we must look for the elimination or reduction of manifest cardiovascular diseases through manipulation of environmental factors acting on the individual. Such manipulation must be concerned with and related to both external and internal environments. The problems associated with such manipulation cannot be described completely at present, but considerations will be made of a few of the more evident factors.

The ultimate development of environmental physiology lies in the development of sophisticated scientific technology's evaluating the interplay between the whole living organism and the environment. Essentially, this means the effective control of the external environment by adjustments of the internal economy of the organism. Adaptability involves the responses of the organism as a whole, above and beyond the reactions of its individual cells or molecules. There must be limits to the range of human adaptability. The determination of these limits and the thresholds of the stimulus levels, as well as the variations in sensitivity following varying degrees of adaptation, present a real difficulty. A serious question must be answered: whether adaptations to the artificial control of the internal environment would lead to dire consequences for the future of man. No matter what phase of life control is studied, there will be unexpected byproducts. As Potter (7) has suggested, further research is required on the concept of optimum stress and on the nature of adaptation in human performance and in the human environment. Adaptability is associated with deadaptation. The human organism appears able to achieve some degree

---

[1] This work was supported in part by Santa Barbara County Heart Association Grant No. R636-8.

of adjustment to any stress to which he has been exposed, but what are the patterns of adjustment when the stress is either removed or ameliorated. Are different mechanisms brought into effect if the response (visible) is delayed in appearance? What are the influences of multiple stress agents on grades of intensity of the stressors? This suggests that the major interests of environmental physiology lie not in the elimination of all stresses, physical and mental, but in defining the irreducible minimum of stress level compatible with maximum performance and optimal level of physical and mental health.

The topic under discussion—i.e., cardiac disease—is narrow only in theory, not in scope. The implications of this broad scope suggest that one should consider at least two areas: social factors (the wide environment) and individual control (the internal environment). The projected population distribution in the 20th century suggests potential decrease in the over-all incidence of heart disease within the next 20 years, since there will be fewer individuals in the susceptible age group. This may lead to a temporary decline in the drive to resolve the problems of heart disease, but this would be disastrous, since the number of such patients should show a marked increase as the youngsters of today reach these critical years within two to three decades. The increase in population density may pose a more serious problem in regard to heart disease. Various surveys have shown that, in spite of such known urban pressures on health as overcrowding, air and water pollution, poor housing, stresses of city transportation, and the accelerated pace of city life, there is no substantial evidence that the over-all health of the urban resident is worse than that of the rural. Yet there are some suggestions that this is not quite the entire story. Urban children and young adults have a higher rate of illness than rural equivalents. Urban children have more days of restricted activity, bed disability, and days lost from school or work. Hutt and Vaisey (3) studied the influence of overcrowding on normal young children. In contrast to work on animals, they found that this situation induced the young children to spend less of their time with others, but they also spent less time on the edge of the crowd trying to escape from it. In the most crowded conditions, they became more noticeably aggressive and destructive. The similarity of these environments to those presently observed to be consistent with the development of cardiac disease is evident. It also implies that a very advanced and complex regulating mechanism is operating in the light of the internal environment. Homeokinetic maladjustments of this regulating mechanism may well lead to the disturbed state most conducive to cardiac disease. Individual members of any population interact with each other. The consequences of this interplay offset many, if not most, physiologic states. How the degree of overcrowding affects hormonal balance, resistance to stress, reproduction, growth, learning, etc., is not known as yet, although some evidence is accumulating to suggest that such mass human involvement does lead to lessened reaction and resistance. All the factors responsible for control of high-density population dynamics are too poorly understood to lead to conclusive statements, but the suggestive hints of disaster are present.

Associated with overcrowding and altered social and work environments is a diminution in physical activity. There has been an increase in the amount of "white collar" work, a decrease in energy expenditure on various jobs, increased participation in spectator activity in sporting events, a diminution in the number of working hours per week, a greater leisure with sporadic episodes of high-intensity physical work, and, in some populations, a marked increase in food intake. Man's capacity for work has not been altered, as evidenced by the increased number of records being broken in sports involving not only skill, but strength and endurance. The capacity is present in those who have become our present-day gladiators, yet there is increasing evidence that man as a group apparently is becoming less physically fit (8). It has also become increasingly clear that, as man ages, not only does his capacity for work decrease, but he is carrying on his activities with a marginal reserve (10). The work tolerance of patients with various diseases is now being studied, but the more important role of environmental factors in either increasing or decreasing the diseased individual's capacity has been hardly considered. Continual and more extensive research in this area should become a primary goal in the field of environmental physiology. Since the incidence of cardiac disease is somewhat less in females, questions as to activity levels, exercise tolerance, the role of female sex hormones, dietary history, metabolic handling of fats, carbohydrates, and proteins, etc., must be related to and considered in the light of potential sex differences. Yet we know very little regarding the capacity of the female versus the male.

Physical activity has been considered to provide benefits toward general health ever since man began to relate the physically fit to the ill individual. It is increasingly evident that we have no basis for or against this concept (1). We cannot determine just how much physical activity is best for the health, not only of the population, but of any one individual. Is exercise, indeed, necessary? If so, when, how, and to what degree should an individual partake or should society enforce a rigid dergee of such activity, regardless of the health of the individual? If the latter be taken into account, as it must, what level of disease is to be associated with how much exercise? Is it better to start progressive programs of exercise at an early age, or can benefits accrue regardless of age or health status when physical activity begins? Just as important a problem as the whens, hows, and whys is the determination of the methods to ensure continued participation in exercise programs if it is proved that they are valuable for prolonging life and maintenance of health.

No adequate data were available relating physical activity to cardiac disease until Morris *et al.* (6) presented their suggestive analysis based on their studies of London transport workers. Although constitutional and psychologic factors may have had influences on the mortality statistics, it was grossly evident that difference in physical activity was a determinant factor. Kahn (4) has reported an excess risk of coronary heart disease of 1.4 to 1.9 times for postal clerks, compared with carriers. He suggests that current physical activity may be more closely associated with the differential than is the individual's lifetime average physical activity on the job. On the other hand, reports (9) without such a positive association are available. Some studies (2) are available documenting the ability of cardiac-disease patients to improve their physical work capacity and even their degree of cardiovascular adaptation by increasing physical activity. However, it would appear that the burden of proof still rests on those who state physical activity has a protective or even preventive influence on cardiac disease (including hypertension). The inconclusive data suggest some possible long-term cardiovascular benefits resulting from increased physical activity. How much benefit accrues, how effectively the benefit is retained, and how much physical activity is needed to reduce and to prevent cardiovascular disease are open questions. This area is and will continue to be one of the most important fields of investigation for environmental physiology.

It has been suggested that coronary thrombosis is a disease of affluence. Affluent societies are associated with a decrease in the physical activity required of its members in the production and utilization of the individual's material needs. Not only has man become more sedentary, but his outlets for physical activity have been reduced to a bare minimum. Consequently, watching, eating, and smoking have become his major preoccupations. While all these factors may be related to the causation of cardiac disease, the dietary factor has undoubtedly aroused the most intense controversy. The hypothesis that the major cause of coronary thrombosis is a high intake of fat, particularly highly saturated fats, has not been established. It is, therefore, not surprising that an equally controversial hypothesis (11), suggesting a causal relationship between high carbohydrate intake (sucrose) and this cardiac disease state, be postulated. These developments in environmental physiology are not unexpected, reflecting as they do the unsure state of the role of environmental factors in the normal and abnormal physiology of man. It is further quite clear that interactions of various elements of the environment have not attained a level of perfection permitting one the joys of decision. All the factors mentioned rather perfunctorily above do no more than suggest the potential impact of many social factors on the adaptation of man to environmental stress and their implication for cardiac disease. The future for investigation is bright but somewhat overwhelming in its magnitude.

The quality of the environment plays an important part in man's interaction and response to environments. In view of some of the marked interferences with this quality, a question must be raised as to whether or not we are attaining a point where man is potentially obsolescent. If the external environment no longer provides adequate stimuli for development or even if it only acts as a potential eliminator of certain essential functions of the human organism, then there seems to be a need to provide systems for regulation of the internal economy through manipulation of control mechanisms and substitution of essential parts.

Developments in substitution therapy of "parts" for individuals with cardiac disease are well on their way. A basic beginning has been made for appropriate substitution of vessels, hearts, kidneys, livers, muscles, joints, etc., and eventual simplification of these, increased availability to all, lower cost, and improved reliability of man-made components are in the immediate future. It is possible that cardiac disease will no longer be a prime factor in man's survival if substitution becomes an accepted part of man's internal structure.

Further developments in substitution of specific "good" genetic material in adults may also lead to a more effective individual in terms of his adaptation to his environment. Although this substitution has been used only in "learning" situations, these experiments raise the possibility that a more proper myocardium may be induced by providing the right DNA for the potential cardiac patient. Similar substitution may be induced for all types of tissues and organ systems.

Probably the most suggestive new environment for man arises from the possibility that programming of control systems will become available. Presently, we have available the cardiac pacemakers, operating rather directly on the target organ; but it is conceivable that, as our knowledge improves, this programming can operate from a higher level of integration. Not only could this be accomplished by implantation of electrodes in the central nervous system, but systems could be designed to provide for autoregulation by proper feedback systems. No longer would the individual have to turn his own dials in anticipation of his needs. Automatic infusion of chemical substances to induce responses, to provide for antagonistic regulation, and to stimulate artificial excretory systems to eliminate specific materials, is a potential possibility for regulation of homeokinetic mechanisms. Blood pressure, cardiac output, and peripheral resistance can be simultaneously regulated to meet the needs of the individual for all and any environmental stress by proper chemical and central nervous system programming.

Although such control of the individual is a distinct possibility, the accomplishment is in the future, depending entirely on the necessary research to clarify all the interactions present. Cardiac disease may become obsolete depending upon not only the above suggestive approaches, but others still in the conceptual stages. Man's control of his environment may be successful, either through internal or external manipulation. Health must be defined and measured in terms of adaptive capacity towards environmental circumstances and hazards. Subjectively, it would seem that, for other aspects of man's success, it would be much better to regulate the external environment and adapt it to his physiologic systems than to modify his internal environment mechanically to meet the stresses of an environment that he has produced and should be able to control and modify for his best interests.

## REFERENCES

1. DAVIES, C. T. M., DRYSDALE, H. C., AND PASSMORE, R. (1963). Does exercise promote health? *Lancet* Vol. II 930–932.
2. DOAN, A. E., PETERSON, D. R., BLACKMON, J. R., AND BRUCE, R. A. (1965). Myocardial ischemia after maximal exercise in healthy men. *Am. Heart J.* 69, 11–21.
3. HUTT, C., AND VAISEY, M. J. (1966). Differential effects of group density on social behavior. *Nature* 209, 1371–1372.
4. KAHN, H. A. (1963). The relationship of reported coronary heart disease mortality to physical activity of work. *Am. J. Public Health* 53, 1058–1067.
5. LINDER, F. E. (1966). The health of the American people. *Sci. Am.* 214, 21–29.
6. MORRIS, J. N., HEADY, J. A., RAFFLE, P. A. B., ROBERTS, C. G., AND PARKS, J. W. (1953). Coronary heart disease and physical activity of work. *Lancet* Vol. II 1053–1057; 1111–1120.
7. POTTER, V. (1964). Society and science. *Science* 146, 1018–1022.
8. RODAHL, K., AND HORVATH, S. M. (1962). "Muscle as a Tissue." McGraw-Hill, New York.

9. Stamler, J., Lindberg, H. A., Berkson, D. M., Shaffer, A., Miller, W., and Poindexter, A. (1960). Prevalence and incidence of coronary heart disease in strata of the labor force of a Chicago industrial corporation. *J. Chronic Diseases* **11**, 405–420.
10. Sullivan, F. J., Bender, A. D., and Horvath, S. M. (1963). The aging cell. *J. Am. Geriat. Soc.* **11**, 923.
11. Yudkin, J. (1967). Sugar and coronary thrombosis. *New Sci.* **33**, 542–543.

# Adaptation and Environmental Control[1]

James D. Hardy and J. A. J. Stolwijk

*John B. Pierce Foundation Laboratory, Yale University School of Medicine, New Haven, Connecticut*

The systems concept of life implies that each unit in the biologic hierarchy is only relatively autonomous and is subordinate to more complex units. A system is a collection of components arranged and interconnected in a definite way so that for a given input there are one or more identifiable outputs. Two types of systems can be recognized: open loop, with no feedback; and closed loop, in which the input to one or more components is affected by the output. Living systems exhibit characteristics of both.

Environmental "stress" is any change in the environment that produces a change or "strain" in the system. At the present time we are restricted to rather gross descriptions of environmental stress, but we can often express quantitatively the deviations from optimal or zero stress conditions.

A system may follow environmental change and thus be termed a "conformer," or it may regulate its state and qualify as a "regulator." Adaptation to environmental heat changes for the homeotherm is not effected by permitting larger changes in internal temperature, but by increasing the capacity of the thermoregulatory system to compensate for the effects of external changes. The process of adaptation is that of the regulator, in changing the limits of its control system, without significantly changing its ouput—internal temperature. It can be argued that a model of a subsystem, such as that of thermoregulation, formulated by itself is unsatisfactory, and that all relevant physiological data should be included, but it needs to be studied in isolation at some stage.

A model has been described for the passive system, i.e., the system to be controlled, in which the head, trunk, and extremities are represented by cylinders with appropriate dimensions to afford conductive and convective heat exchange with the environment. The heat flow relationships can be described in the form of eight simultaneous differential equations, and a number of algebraic expressions and conditions. It was tested by comparisons of predicted changes with those of subjects completely immersed in a shallow bath. The agreement between theoretical and experimental results encourages the belief that the model is sound for these conditions, and that the introduction of a central blood compartment as a center of convective heat transfer is an improvement on earlier models.

The thermostatic regulator accepts signals from the passive system and exerts corrective actions on that system when the signals indicate a departure from preferred conditions. Quantitative data on effector actions are available, but our

[1] Synopsis of presentation, the substance of which has been published elsewhere.

knowledge of the sensor mechanisms is incomplete. For model building purposes they are assumed to be in the hypothalamus, the skin, and the muscles. It is assumed that the signals from the sensors vary linearly with the degree of departure from the set point, that their output is zero at the set point, and that the hypothalamic temperature determines whether the system is switched to prevent overheating or undercooling. The set points have been estimated to be 36.60°C for the hypothalamus, 34.10°C for the skin, and 35.88°C for the muscles. In the model, deviations of hypothalamic temperature are multiplied by the deviations in skin temperature and muscle temperature. This gives a reasonably good simulation of observed hysteresis type responses.

The model has been tested against observed bodily responses by the "ice cream" experiment, in which a measured quantity is given to a subject in equilibrium and the opposing biphasic changes in tympanic and skin temperature are followed. The model predicts the results of the experiment quite well.

The study of thermoregulation is reaching the stage in which descriptive study will yield progressively less and the quantitative and analytical approaches become increasingly necessary. An analog model can be of considerable assistance in both the planning and the interpretation of experimental work. For example, the above model explains the irregularity seen in the shivering induced by cold air as opposed to the steady shivering from immersion in cold water.

In adaptation to heat, at least two effects are seen: (1) a shift in the body temperature level at which sweating is initiated; and (2) an increase in the slope of the sweat response curve. Both effects can be explained by an increase in the responsiveness of the controlling element for sweat production or of the feedback elements in the acclimatized man. The action is upon the controlling system rather than upon the passive or controlled system.

The concepts reported at the Symposium are described in detail in the following publications:

STOLWIJK, J. A. J., AND HARDY, J. D. (1966). Temperature regulation in man—a theoretical study. *Pflügers Archiv.* **291**, 129–162.
HARDY, J. D. (1967). Adaptation to physical stresses. *In* "Cecil-Loeb Textbook of Medicine" (Paul B. Beeson and Walsh McDermott, eds.), 12th ed. Saunders, Philadelphia.
GAGGE, A. P., RAPP, G. M., AND HARDY, J. D. (1967). The effective radiant field and operative temperature necessary for comfort with radiant heating. *ASHRAE*, May, 1967 (Condensed Technical Paper).
GAGGE, A. P., STOLWIJK, J. A. J., AND HARDY, J. D. (1967). Comfort and thermal sensations and associated physiological responses at various ambient temperatures. *Environ. Res.* **1**, 1–20.
GAGGE, A. P., AND HARDY, J. D. (1967). Thermal radiation exchange of the human by partitional calorimetry. *J. Appl. Physiol.* **23**, 248–258.

# Review and Comment on "Waste Management and Control"— A Report to the Federal Council for Science and Technology[1]

DONALD H. PACK

*Air Resources Laboratories, Environmental Science Services Administration, Silver Spring, Maryland 20910*

President Nixon is marshalling the resources of the United States in a program for environmental quality. The Congress, students, industrialists, and the public at large are all uniting to consider the problems of the environment and formulate decisions and action programs.

Four years have passed since "Waste Management and Control" was published; nevertheless, it remains a thoughtful and often provocative distillation of a variety of ideas and opinions representing a broad view of interrelations in resource systems. It is well worth reading, or reading again.

In the report, pollution was defined as "an undesirable change in the physical, chemical, or biological characteristics of our air, land, and water that may or will harmfully affect human life or that of other desirable species, our industrial processes, living conditions and cultural assets, or that may or will waste or deteriorate our raw material resources."

Conservation, one of the two main themes, was defined to include the wasteful effects of pollution ranging from lost work to the wastage of human life. Wastage of resources includes vegetation damage and losses in agricultural yields. Less direct wastage occurs through the denial of waters for recreation, loss of esthetic views, and the sense of impossibility of escape from the displeasing scene. Economic as well as biologic man is considered. Affluence has developed at the price of pollution; affluence can correct it.

Integrated analysis, the second main theme, requires that we balance the cost of desirable goods and services against the cost of reducing the unpleasant and harmful effects of their production. The art of politics and management must take up the burden of translating research findings into actual environmental improvement. The mandatory requirement for viewing the problem in all its aspects requires a systems approach. A correct systems analysis leads to optimization; which implies that certain areas or groups will not realize all their objectives. Optimum must be defined and goals set.

Four sets of problems emerge as especially important and urgent: documentation of pollution levels; determination of the effects of pollution; development of technology for recycling; and innovation of management methods. The measurement of effects is difficult. The great need is to identify quantitatively the portion of the stress created by pollution alone and the additional burden upon

[1] National Academy of Sciences—National Research Council, Publication 1400, 1966 (synopsis of presentation not otherwise published).

individuals already subjected to other stresses. Assessment of the effects of pollutants alone and in combination is a tremendous task, made greater by the variability of the population at risk. Which levels should be chosen to protect what group? Should it be the most sensitive? When, as here, the evidence is not clear, or the effects are subtle, and the cost of correction is high, the tendency is to pretend that the problem does not exist.

Since it is not now known what concentrations result in deleterious effects, the measurement program tends to use the most sensitive measure that the state of the art and finance permit. More efficient measures should come with increased knowledge of effects. The report stresses the need to determine effects, to set monitoring needs, to assist in developing criteria and standards, and to estimate costs in reaching these goals.

The report describes and deplores the present wastage of resources and despoilation of the environment; it is apprehensive of the future; but it encourages boldness and imagination in the attack on environmental pollution.

## How Is an Optimum Environment Defined?

VAN R. POTTER

Any discussion of the future in the context of physiologic hazards in man's environment will inevitably have to be broad enough, it seems to me, to consider leisure as a possible health hazard, and at the other extreme, information overload (15) or future shock (25) as hazards which society will have to consider. A full-page advertisement that appeared in 1958 illustrates what I mean by the hazards of leisure. It is entitled, "Power companies build for your future electric living," and subtitled, "Electricity may do your yard work." I think it was this that first aroused my interest in the physiologic hazards of leisure, a commodity that you and I have not suffered from particularly, if we define "leisure" as time that provides relief from responsibilities.

The major message that I wish to deliver is that an optimum environment is one that delivers what I would like to call an "optimum stress;" but I have come to realize that the word "stress" produces stress in many physiologists, and therefore I will not again refer to "optimum stress." Instead, I will follow the definitive "Perspectives of Adaptation" presented by Prosser and by Adolph in Sect. 4 of the Handbook of Physiology: "Adaptation to the Environment" (7). Prosser defines a "stressor" as an environmental circumstance which induces adaptations, and what I shall be talking about is the concept of deliberately promoting an optimum stressor level (Fig. 1). Indeed, I wish to promote the idea of a merger between physiology, genetics, and molecular biology on the one hand, with anthropology, psychology, and philosophy on the other. Our goal for the future should be to adapt our evolving culture to the fact that adaptation is a normal feature of life and that we need to know much more about the process than we do now. But it is not enough to consider adaptation as an esoteric topic dealt with by physiologists. It is a topic that is fundamental to all the specialties mentioned above, and moreover it is a topic that needs to be incorporated into our educational system and into our cultural milieu.

### CULTURE AS ENVIRONMENT

Prof. Clifford Geertz (10) of the Anthropology Department of the University of Chicago has contributed an excellent chapter to the book entitled "New Views on the Nature of Man," edited by John R. Platt. It is entitled "The Impact of the Concept of Culture on the Concept of Man," and its thesis is that (1) culture is a set of control mechanisms for controlling or programming human behavior, and (2) man is precisely the animal most desperately dependent upon such extragenetic, outside-the-skin control mechanism for ordering his behavior. He says: "One of the most significant facts about us may finally be that we all begin with

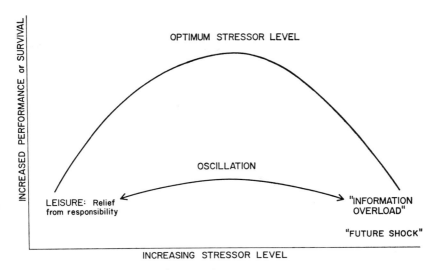

FIG. 1. Definition of optimum environment.

the natural equipment to lead a thousand lives but end in the end having lived only one."

The question is whether we, as physiologists, molecular biologists, and geneticists, are prepared to tell society how to move toward a cultural tradition that can help each individual find an optimum stressor level and steer successfully between the Charybdis of too much leisure and the Scylla of information overload.

The question I am raising is whether our 20th-century culture is providing the proper kind of "extragenetic, outside-the-skin control mechanisms" that the human animal is so "desperately dependent upon." Do prospective parents know that every individual is genetically different from every other individual, that only those genetic capabilities that are possessed can ever be expressed, and that every genetic capability can be fully expressed only in the presence of an optimum stressor level, applied at the right time—and then not continuously? Do we, as scientists and humanists, have any idea what to tell these hypothetical prospective parents if they ask us for practical answers to these questions?

There are historical precedents for these questions. In 1890, we find William James saying, "Keep the faculty of effort alive in you by a little gratuitous exercise every day. That is, be systematically ascetic or heroic in little unnecessary points, do every day or two something for no other reason than that you would rather not do it, so that when the hour of dire need draws nigh, it may find you not unnerved and untrained to stand the test," (2, p. 715). And in 1919, we hear the 20th-century cynicism replying, in the words of Somerset Maugham: "I forget who it was that recommended men for their soul's good to do each day two things they disliked: . . . it is a precept that I have followed scrupulously; for every day I have got up and I have gone to bed" (2, p. 875). In the same vein, I believe it was Robert Maynard Hutchins who spoke for academicians in general, when he said: "Whenever I feel the urge to exercise coming on, I lie down until the feeling goes away."

## PHYSIOLOGIC ADAPTATION

What Maugham and Hutchins do not seem to realize is that adaptation is not something that is just a physiologic response to some abnormal or novel situation, some hypoxic condition at a mountain top where average people never go. The living animal is adapting during every moment of life. Adaptation is not merely a mechanism "for the adjustment of organisms to the action of unfavorable (sic) factors in the environment" (Barbashova in Ref. 7, p. 49). Adaptations are part of the everyday life mechanism, and they are noted as adaptations by the physiologist only when they differ from "ordinary responses" sufficiently to be "increments or decrements in the ordinary response" that "require time to develop" (Adolph in Ref. 7, p. 34). The word "adaptation" is frequently misunderstood because it has come to have two meanings. Some people think of evolutionary adaptation, which involves natural selection of mutants that fit a particular environment. This kind of adaptation occurs in populations over a large number of generations. The other kind of adaptation is physiologic adaptation, which applies to the individual members of a population.

In the present discussion, I am concerned only with physiologic adaptation and follow Prosser (in Ref. 7, p. 11), who defines it as "any property of an organism which favors survival in a specific environment," although Prosser adds the words that narrow the meaning when he says "particularly a stressful one." Prosser states that "an adaptation permits maintenance of physiological activity and survival when the environment alters with respect to one or more parameters."

As a biochemist and molecular biologist, I have the thesis that, if we examine the organism at the molecular level, the environment is changing on a minute-to-minute basis and the "ordinary responses" at the enzyme or nucleic acid level constantly are undergoing adaptive increments or decrements in enzyme activity or amount in a complex feedback system that is designed to maintain certain features of the internal milieu within limits that will promote survival. We can either cultivate this adaptive capacity or we can let it atrophy.

The key point is that, "if an organism can adapt, by the same token it will 'de-adapt.' When we carry a heavy load, our ability to carry loads increases, but the corollary is that, when we have no loads to carry, our ability to carry loads decreases. Just how big a load should we be able to carry, and what kind of load-carrying program will be best for each individual?" In this context, a load can, of course, be physical, intellectual, or emotional (7). The idea that an effort should be made to define "health" as more than the absence of disease was emphasized by Rene Dubos in "Man Adapting" (8), when he said (p. 391): "A medical philosophy which assumes *a priori* that the ideal is to make human life absolutely safe and effortless may paradoxically create a state of affairs in which mankind will progressively lose the ability to meet the real experience of life and to overcome the stresses and risks that this experience inevitably entails." I think that Dubos would be willing to include individual men, as well as mankind in the above statement.

Elsewhere, Dubos comments (pp. 362–363) that "in the United States, the emphasis is on controlling disease rather than on living more wisely." "The public health services themselves, despite their misleading name, are concerned less with health than with the control of the specific diseases considered important in the communities they serve. They do little to define, recognize, or measure the healthy state, let alone the hypothetical condition designated 'positive health.'"

I suspect that most of us feel that the name of the public health service is not so misleading, and that the evolution of concepts will naturally move toward the concept of positive health as infectious diseases and malnutrition are conquered and degenerative diseases continue to increase. In this event, we must not only eliminate dangerous chemicals from our environment as much as possible, but also learn how to live more wisely.

## ENZYME ADAPTATION TO TOXIC HAZARDS

Physiologic adaptation at the enzyme level is a subject that is close to the core of this entire conference, as well as to the core of this session. Should we aim for an environment that is completely free from toxic hazards? We can answer that question very quickly by recognizing that such an achievement is not only improbable, it is impossible! Then what is the effect of a low dose of a toxic substance, a dose that does not produce death or obvious pathology in a time that is short enough to suggest an obvious cause-and-effect relationship? The answer is that, for many toxic substances at dosages that produce no rapid overt symptoms in the intact animal, there is a rapid increase in activity of a variety of enzymes within 24 hours, which appears to be a result of synthesis of new enzyme, first indicated by the experiments of Conney, Miller, and Miller (6). Numerous examples are now available (9).

So the next question is whether the animal is any worse off or any better off for having an increased level of certain enzymes in liver (6, 9), or intestine (26). In the study of Wattenberg and Leong (26), it was shown that 6 mg of phenothiazine increased the activity of benzpyrene hydroxylase in rat liver from 14 to 311 units/mg, and in intestinal mucosa from 16 to 55 units/mg. Chlorpromazine, a phenothiazine derivative that is a well-known tranquilizer, had almost identical results for liver, and increased the value in intestine to 114 units. At a dose of only 1 mg per rat given 48 hours prior to exposure to 30 mg of dimethylbenzanthracene, phenothiazine protected nine of ten rats against adrenal necrosis. This study has as its over-all objective the determination of "optimum methods for achieving protection from the effects of chemical carcinogens in experimental animals and, possibly, eventually in man." Further studies on the protective effects of smaller daily doses, such as are used over long periods of time as tranquilizers, will be of interest.

Meanwhile, a clinical application of the induction of enzyme activity in a human has come to our attention. Yaffe, Levy, and Matsuzawa (27) have reported the use of phenobarbital (15 mg daily) to enhance glucuronide conjugation capacity in a congenitally hyperbilirubinemic infant. The authors claim that theirs is the first demonstration of the apparent induction of the glucuronide conjugating system by phenobarbital in man, and the first therapeutic application of enzyme induction. There is no reason to doubt that phenobarbital did cause increased synthesis of the enzyme, since enzyme induction by phenobarbital is well known (9).

At the present moment, it appears impossible to answer the question as to whether amounts of toxic substances that will induce certain enzymes are intrinsically helpful or harmful. Of course, we have been made aware during this conference that even noninducing doses of some compounds may be capable of producing irreversible damage to DNA, and hence capable of acting as carcinogens. Recently, Siperstein (23) has reported deletion of the negative feedback system for cholesterol biosynthesis in liver by aflatoxin at a level of 10 parts per billion in the diet of trout and in the liver of rats after only 2 days of a diet containing aflatoxin at a level of 10 ppm. In this case, it is not yet established whether the change is irreversible or reversible, but it is clear that the phenomenon could give the appearance of enzyme activation or induction. It should be noted that trout receiving aflatoxin at 10 parts per billion from birth develop hepatomas to the extent of 80%–100%. Thus, we are in the position of having to make considerable efforts to find any virtue in the ingestion of any foreign compound if it can be avoided, unless there is clear indication that the benefits outweigh the possible or known risk. Further discussions on animals in toxic environments are available in the Handbook referred to earlier (7). No beneficial effects seem to be mentioned in the section on polluted air (by E. B. Morrow, on pp. 795–808 in Ref. 7) or in the section on man and industrial chemicals (by H. W. Gerarde, on pp. 829–834 in Ref. 7), although in both chapters there are suggestions of adaptation in various instances. Tables showing levels to which workers can be exposed day after day without adverse effect do not appear to have been based on criteria other than the ability to work efficiently and the absence of claims of pathology. Thus, at the moment, we can suggest that, although there is evidence that humans can adapt to some extent to a variety of toxic chemicals there is still no basis for assuming that there are no disadvantages to the adapting individual, and no basis for complacency about the increasing levels of ambient toxicity. Meanwhile, let us examine the type of adaptive changes that might be considered in examining the life of a completely healthy man or animal.

## ENZYME ADAPTATION TO DAILY REGIMENS

Changes in enzyme amount resulting from changing rates of synthesis and degradation of enzyme protein occur not only in physiologic adaptation to toxic hazards in the environment. Such changes can also be observed in the course of each 24-hour cycle, as a result of feeding habits and as a result of changes in dietary composition. In 1961, Ono and I (21) reported enzyme changes in liver rats fed *ad libitum* under controlled lighting conditions, in which food was eaten in the dark hours. The liver glycogen decreased in the light hours and increased in the dark hours, and glucose 6-phosphate dehydrogenase showed mild oscillations in opposite phase. More recently, we (20) studied a number of enzymes in livers and hepatomas of rats fed diets containing 0%, 12%, 30%, 60%, and 90% protein only during 12 hours of darkness with no food during 12 hours of light in each 24. We noted oscillations in liver glycogen with both peak and minimum levels that were inversely related to protein and directly proportional to carbohydrate in the diet. In this study, there was a marked oscillation in tyrosine transaminase in rat liver, with maximum values proportional to protein intake at midnight, 6 hours after the onset of feeding, with declines to resting levels thereafter, even while the glycogen values were still rising. Recently, this phenomenon has been studied in more detail by Dr. Minro Watanabe in our laboratory, and the detailed data confirm the discontinuity between liver glycogen and liver tyrosine transaminase, and in addition show that the plasma corticosterone decreases while the rise in tyrosine transaminase occurs. The induction of this enzyme by injections of adrenal steroids was well known, but the rapid increases during food ingestion have not been previously reported, possibly because in the rat the peak is at midnight and tends to reach daytime levels by 6 AM (28–31).

These examples show that metabolic transitions occur throughout the daily routine of sleep and waking, eating and not eating, and that these transitions are accompanied by marked changes in several enzyme activities.

## ENZYME ADAPTATION TO FASTING

In the 1961 study (21), we also carried out the procedure of fasting rats for 3 days and then refeeding various diets, as has been carried out by Tepperman and Tepperman (24). In this situation, the glucose 6-phosphate dehydrogenase activity in the liver responds by rising to unusually high levels, compared with the daily cycle owing to a changed feedback relationship, and greatly overshoots the normal range. The enzyme level rises to about 20 times the usual level during about 72 hours, and then slowly returns to normal levels over a period of about 10 days. It is clear that the optimum stressor level for the synthesis of this enzyme is not attained in the *ad libitum* feeding regimen.

In a recent symposium on feeding patterns and their consequences, Mayer (14) emphasized that the pattern of metabolism may be influenced by the distribution of the food intake in one, two, three or a number of meals. Tepperman and Tepperman (24) and others have carried out many experiments on rats that have been fasted for 21 or 22 hours in each 24, and correspondingly fed for only 2 or 3 hours per day. Such animals showed a striking increase in fat production from both acetate and glucose, and they stored liver glycogen to higher levels than animals fed *ad libitum*. Moreover, the fat pads showed an increased level of fat deposition. From these studies, there has been a tendency to conclude that frequent snacks are better than one meal a day for the human, as well as the rat or mouse, and there is a widespread belief that frequent small meals represent a regimen that can be recommended as one feature of an optimum environment. It is my purpose here to question this point of view and in fact to take the opposite position. Incidentally, it is reported that the custom of three full meals a day has been established only since 1890; that in Anglo-Saxon tradition there were only two meals a day, breakfast and dinner; and in the 16th century dinner was eaten at 11 AM (4).

We start from the assumption that the animal or human is not equipped by instinct to do what is best for him in circumstances other than those under which his species evolved to its present state. Particularly for rats in small cages or men in small rooms, with no exercise imposed by circumstances, the provision of food on an *ad libitum* basis has no instinctive support to provide information on "living more wisely," a goal emphasized by Dubos (8), as mentioned earlier. True, our society has come to value the slim figure, achieved by a low-calorie diet, but this has been coupled with the idea of frequent snacks and never, to my knowledge, with the idea of one good meal once a day.

Secondly, we start from the assumption that fasting is normal, beneficial, and not unnatural. This view is probably contrary to popular and medical opinion, both of which are probably influenced by the knowledge that stomach ulcers are given palliative treatment by frequent small meals of milk. My assumption is that physiologic functions are meant to be used, and that atrophy of disuse is not limited to skeletal muscle. It is also assumed that brain function, both intellectual and characterwise, is influenced by bodily function. Animals possess metabolic machinery for storing fuel far in excess of what can be stored in the form of liver glycogen. According to Cahill (5), the liver glycogen in man is able to maintain blood glucose for only about 12 hours, and, of course, muscle glycogen is not directly available as a source of glucose. Thus, in the absence of food intake, the human must initiate the process of gluconeogenesis from protein within 12 hours, and the evidence suggests that in rodents the situation is not greatly different. Cahill shows the relative sizes of the various glucose reserves in comparison with the total potential reserves, and the total glucose pool is only about 290 moles, or a little less than 1% of the total of about 30,000 moles, based on the fact that men can survive a total fast of 60–90 days.

It has commonly been assumed from studies on laboratory rats that fasting leads to a rather prompt adrenal cortical stimulation, which in turn leads to increased gluconeogenesis from protein (1, 22). However, in a recent study by Kollar *et al* (13), two men were fasted 4 days and three men 6 days, with no evidence of adrenal activation in any of them. However, an as yet unidentified "stress responsive indole substance," normally absent from the urine, was maintained at high levels in all six subjects after the second day of starvation. What lends interest to this observation is the finding by Rosen and Nichol (22) that two tryptophan metabolites, 5-OH-tryptophan and serotonin, are effective inducers of tyrosine transaminase in rat liver. It appears that the stress of fasting may lead to adaptive changes mediated both by indole-containing substances and by cortisone and that the two classes of inducers are merely separate stages in a sequence of adaptive changes, which vary in different species.

The interest in my laboratory has not been in the effect of fasting *per se*, but rather in the effect of prolonging the intervals between food intake as an approach to an optimum stressor level. The typical laboratory rat lives at a constant temperature, with food constantly available, and with little opportunity or incentive to exercise. In rats that have been trained to eat their daily allotment of food in 2 or 3 hours the result is obesity (24). Under natural selection, rats have been selected to desire food every night and to sleep in the daylight hours, but they have also been selected for the ability to survive if they do not obtain food every night. We decided to explore the consequences of training rats to undergo cycles of feeding and fasting with durations of 48 hours, rather than 24 hours.

From the experience with *ad libitum* feeding regimens, with food limited to 12 hours in each 24, and with food *ad libitum* after a 3-day fast, we went on to develop regimens that have attempted to attain the hypothetical optimum stressor levels corresponding to an environment that might be optimum for the greatest number of parameters. Starting with rats that were fed 12 hours in each 24, which is not much different from *ad libitum* feeding, we adopted the simple expedient of feeding every other night, instead of every night. Thus these animals are fed 12 hours in every 48, and are fasting for 36 hours. We refer to them as "36-hour fasting-adapted rats" (19). Some experiments have been done with a 48-hour cycle in which food is available for only 8 hours in 48, and these are called "40-hour fasting adapted rats" (28–31).

It was soon noticed that the fasting-adapted rats are hyperactive. We therefore installed activity cages and recorded mileage for each animal. It can be reported that 40-hour fasting-adapted rats will survive in small cages of the usual type, but if allowed access to activity cages they will all run themselves to death in 5–10 days, illustrating the breakdown of instinctive "wisdom." In contrast, the 36-hour fasting-adapted rats are able to survive and adapt to the regimen with or without exercise.

The 36-hour fasting-adapted animals run an average of 15 km during the 12-hours of darkness when no food is available (19), and on the following day, still hungry, they sleep much of the time and run much less. The *ad libitum* controls exercise and eat at night and sleep most of the day, with almost no day activity recorded.

We have studied a number of enzymes and have also studied nucleic acid metabolism, but data for only one enzyme will be mentioned here. Glucose 6-phosphate dehydrogenase shows great fluctuations and averages much higher values than the *ad libitum* controls (19). We feel that these data adequately demonstrate the fact that adaptive enzyme changes are part of normal physiology and that these changes can be exaggerated by increasing the time between meals.

The animals on this regimen have not been studied for longevity, but it is clear that obesity does not result during 30 days, and that the growth curve during this period does not develop a steeper slope than the control curve, as in the case of rats fed 2 hours per day (12). The growth rate is quite similar to that of male rats studied by Berg (3) at the levels of 33% and 46% restriction of food intake, and falls midway between the rates for his groups. Berg's studies demonstrated increased longevity and general health without evidence of immaturity. In the 36-hour fasted rats, it is as yet unknown whether the food intake would increase to the level of the *ad libitum* feeding in time, or whether in the long run restricted feeding on a once-per-day basis would be preferable. It can be suggested, however, that the usual restricted food intake always results in fasting periods of 22–23 hours, and that fasting may be partly responsible for the beneficial effect of the caloric restriction.

At this time, we can only speculate that the fasting-adapted animals derive benefit from the active use of their abilities to produce large shifts in intracellular enzyme patterns, and leave it to the future to determine whether mental powers, motivations, and character would be helped or hindered in humans on similar regimens.

## DEFINITION OF OPTIMUM ENVIRONMENT

Turning now to the actual task of defining an optimum environment, I think I have tried to make the point that the culture in which we live is an important part of our environment. What we recommend as physiologists may become a part of our culture or it may be ignored, depending on the force of our logic and the force of counterpressures. A cogent example is the relationship between cigarettes and lung cancer, where the experts are pretty much agreed but the counterforces still prevail.

In the case of defining an optimum environment, I think that it is essential to go beyond a mere consideration of toxic hazards and to recommend positive steps that could raise men above mere absence of disease to the concept of "positive health" mentioned by Dubos (8). My basic concern is that physiologists should be in the forefront of describing and analyzing the mechanisms of physiologic adaptation, not only in physiologic terms, but in behavioristic and molecular terms—not only in the academic sense, but also with the aim of guiding mankind during the next 20 or 30 years of rapid change to we know not what. If affluence is not conducive to an optimum environment, then what can we say to those who seek it? What are the minimum requirements for a world that physiologists could recommend and work toward?

I have made a list of seven points, not as a physiologist but as a scientist and a humanist. I present them as the whole package, recognizing that, as I am a physiologist, my main concern is the fourth point, which has to do with physiologic adaptation. However, I might emphasize that, until we know more about the nature of adaptation and the molecular targets of toxic hazards, we can say very little about the problems of threshold and potentiation, which are the crux of the pollution problem.

My seven points are as follows:

1. I would begin with basic needs that can be satisfied by effort. These include food, shelter, clothing, space, privacy, leisure, and education, both moral and intellectual. (Regarding the ethical basis of science, see Glass (11).

2. I would insist on freedom from toxic chemicals, unnecessary trauma (primarily war and traffic injuries), and preventable disease.

3. I would demand a culture that had a respect for sound ecologic principles with a long-range point of view. The aim must be to live with nature as in a balanced aquarium, with no assumptions about the ability of future science to rescue a sick planet.

4. The point that is the most personal with me, as it is based on my research experience over many years, is that the culture should prepare us for and expect of each individual adaptive responses, continually from birth to death, as a result of systematic challenges by physical and mental tasks which come at appropriate times, and are within individual capabilities, which are known to

increase rapidly at certain periods in life and decline later. This proposition is based on molecular genetics, which shows that our genetic capacity does not automatically express itself, and that each capability is called forth maximally only in response to an optimum stressor level.

5. The next point calls for individual happiness that involves oscillation between satisfaction and dissatisfaction, with a sustained sense of identity, despite a continual revision of ongoing intentions as the new messages from a rapidly changing world show up in our private information centers.

6. The next point calls for productivity that involves commitment to other members of society, directly to our family, our church, our community, or indirectly to mankind.

7. Finally, each of us can contribute to the further evolution of our culture, to our present society, and to our own satisfaction by a continual search for beauty and order that does not deny the role of individuality and disorder (16, 17).

In connection with the subject of adaptive responses, which is my main concern under my assigned topic of how to define an optimum environment, I feel it is within the scope of environmental physiology (*a*) to postulate that individual organisms show a wide variation in their tendency to seek out such challenges unaided, and that most individuals in fact are equipped by instinct and natural selection to satisfy only their basic physiologic needs by the easiest possible route and no more; (*b*) to inquire what counsel to give a society that is basically uninformed as to the nature of man as well as to the goals, attitudes, and personal values which should be inculcated into preschool and in-school individuals in order to build a free society worthy of mankind; and (*c*) to inquire whether individuals are in general unwilling to challenge systematically their own adaptive capabilities and whether society should attempt to influence the attitude of the young in favor of systematic exercise, periodic fasting, and continual development of new mental and motor skills, and, if so, how these adaptive challenges can be initiated by society but maintained by free individuals, perhaps reinforced by some kind of voluntary grouping.

The first step in any program that might hopefully assist in the evolution of our culture toward better concepts of positive health is to secure more knowledge. I wish to advocate that the study of the individualistic aspects, as well as the principles, of physiologic adaptation is an especially important sector of environmental physiology and that much more research which combines both organismic and molecular biology should be our first objective. Such studies should in particular be cognizant of the biologic rhythms stressed by Aschoff and the individual variations stressed by Sargent at these sessions.

## REFERENCES

1. ASHMORE, J., WAGLE, S. R., AND UETE, T. (1964). Studies on gluconeogenesis. *Advan. Enzyme Reg.* **2**, 101–114.
2. BARTLETT, J. "Familiar Quotations." 13th and Centennial Edition. Little, Brown, Boston, Massachusetts (1955).
3. BERG, B. N. (1960). Nutrition and longevity in the rat. I. Food intake in relation to size, health and fertility. *J. Nutr.* **71**, 242–254.

4. BRASCH, R. "How Did It Begin?" McKay (1966).
5. CAHILL, G. F., JR. (1964). Some observations on hypoglycemia in man. *Advan. Enzyme Reg.* **2**, 137–148.
6. CONNEY, A. H., MILLER, E. C., AND MILLER, J. A. (1956). The metabolism of methylated aminoazo dyes. V. Evidence for induction of enzyme synthesis in the rat by 3-methylcholanthrene. *Cancer Res.* **16**, 450–459.
7. DILL, D. B. "Handbook of Physiology." Sect. 4. *Am. Physiol. Soc.* Williams & Wilkins, Baltimore, Maryland (1964).
8. DUBOS, R. "Man Adapting." Yale Univ. Press, New Haven, Connecticut (1965).
9. ERNSTER, L., AND ORRENIUS, S. (1965). Substrate-induced synthesis of the hydroxylating enzyme system of rat liver microsomes. *Federation Proc.* **24**, 1190–1199.
10. GEERTZ, C. The impact of the concept of culture on the concept of man. In "New Views on the Nature of Man" (John R. Platt, ed.). Univ. of Chicago Press, Chicago, Illinois (1965).
11. GLASS, B. (1965). The ethics of science. *Science* **150**, 1254–1261.
12. HOLLIFIELD, G., AND PARSON, W. (1962). Metabolic adaptations to a "stuff and starve" feeding program. II. Obesity and the persistence of adaptive changes in adipose tissue and liver occurring in rats limited to a short daily feeding period. *J. Clin. Invest.* **41**, 250–253.
13. KOLLAR, E. J., SLATER, G. R., PALMER, J. O., DOCTER, R. F., AND MANDEL, A. J. (1964). Measurement of stress in fasting man. *Arch. Gen. Psychiat.* **11**, 113–125.
14. MAYER, J. (1966). Introductory remarks. Symposium on Feeding Patterns and their Biochemical Consequences. *Federation Proc.* **23**, 59.
15. MILLER, J. G. Psychological aspects of communication overloads, pp. 201–204. Intern. Psych. Clinics: Communication in Clinical Practice (R. W. Waggoner and D. J. Carek, eds.). Little, Brown, Boston Massachusetts (1964). Other reports by Miller include: Adjusting to Overloads of Information, pp. 87–100. In "Disorders of Communication." XLII: Research Publications, Assoc. Res. in Nervous & Mental Disease, 1964; and The individual as an information processing system. pp. 1–28, In "Information Storage and Neural Control" (W. S. Fields and W. Abbott, eds.). Thomas, Springfield, Illinois (1963).
16. NEUTRA, R. S. "Survival Through Design." Oxford Univ. Press, New York (1954).
17. POTTER, V. R. (1964). Society and science. *Science* **146**, 1018.
18. POTTER, V. R., AND AUERBACH, V. H. (1959). Adaptive enzymes and feedback mechanisms. *Lab. Invest.* **8**, 495–509.
19. POTTER, V. R., GEBERT, R. A., AND PITOT, H. C. (1966). Enzyme levels in rats adapted to 36-hour fasting. *Advan. Enzyme Reg.* **4**, 247–265.
20. POTTER, V. R., GEBERT, R. A., PITOT, H. C., PERAINO, C., LAMAR, C., JR., LESHER, S., AND MORRIS, H. P. (1966). Systematic oscillations in metabolic activity in rat liver and in hepatomas. I. Morris hepatoma 7793. *Cancer Res.* **26**, 1547–1560.
21. POTTER, V. R., AND ONO, T. (1961). Enzyme patterns in rat liver and Morris hepatoma 5123 during metabolic transitions. *Cold Spring Harbor Symp. Quant. Biol.* **26**, 335–362.
22. ROSEN, F., AND NICHOL, C. A. (1964). Studies on the nature and specificity of the induction of several adaptive enzymes responsive to cortisol. *Advan. Enzyme Reg.* **2**, 115–135.
23. SIPERSTEIN, M. D. (1966). Deletion of the cholesterol negative feedback system in precancerous liver. Abst. *J. Clin. Invest.* **45**, 1073.
24. TEPPERMAN, H. M., AND TEPPERMAN, J. (1964). Adaptive hyperlipogenesis, *Federation Proc.* **23**, 73–75.
25. TOFFLER, A. (1965). The future as a way of life. *Horizon* **7**, No. 3, 108–115.
26. WATTENBERG, L. W., AND LEONG, J. L. (1965). Effects of phenothiazines on protective systems against polycyclic hydrocarbons. *Cancer Res.* **25**, 365–370.
27. YAFFE, S. J., LEVY, G., MATSUZAWA, T., AND BALIAH, T. (1966). Enhancement of glucuronide-conjugating capacity in a hyperbilinemic infant due to apparent enzyme induction by phenobarbital. *New Eng. J. Med.* **275**, 1461–1466.

## ADDITIONAL REFERENCES

The following references were received after the manuscript was prepared:
28. BARIL, E. F., AND POTTER, V. R. (1968). Systematic oscillations of amino acid transport in liver from rats adapted to controlled feeding schedules. *J. Nutr.* **95**, 228–237.
29. POTTER, V. R., BARIL, E. F., WATANABE, M., AND WHITTLE, E. D. (1968). Systematic oscillations in liver from rats adapted to controlled feeding schedules. *Federation Proc.* **27**, 1238–1245.
30. WATANABE, M., POTTER, V. R., AND PITOT, H. C. (1968). Systematic oscillations in tyrosine transaminase and other metabolic functions in liver of normal and adrenalectomixed rats on controlled feeding schedules. *J. Nutr.* **95**, 207–227.
31. WHITTLE, E. D., AND POTTER, V. R. (1968). Systematic oscillations in the metabolism of orotic acid in the rat adapted to a controlled feeding schedule. *J. Nutr.* **95**, 238–246.

# INDEX

## A

Absorption, species differences in, 37
Acceleration, 171
  cross-adaptation and, 162
Acclimation, see also Cross-adaptation
  definition of, 89
  types of, 91–93
Acclimatization, definition of, 89
Acetanilide, toxicity and, 46
Acetate, 221
N-Acetyl-4-aminoantipyrine, central nervous system penetration by, 11
2-Acetylaminofluorane, 46
Acetylaminofluorene, carcinogenesis and, 65, 71
Acetylcholine
  bronchial response and, 80–81, 82, 83
  parathion and, 17
S-Acetylglutathione, 59
Acids
  central nervous system penetration by, 11
  organic, red cell penetration by, 10
ACTH, see Adrenocorticotropic hormone
Actinomycin D, 38
  aflatoxin and, 53
Adaptation, 4, see also Cross-adaptation
  aging and, 125
  cardiac disease and, 206–210
  chemical, 170, 171
  cold and, 202–203
  cultural, 149–153, 154
  definition of, 88, 170
  enzymes and, 219–224
  genetic, 101–102
  health and, 155
  heat and, 177–202, 203, 205, 212–213
    behavioral and EEG changes and, 188–190
    chronic exposure to heat and, 190–193
    ethnic differences and, 183–188
    individual variability and, 177–183
    performance of tasks and, 193–200
  interaction of organisms and, 122–123
  man and, 95–99
  molecular mechanisms of, 93–95
  optimum environment and, 216–225
  physiological, 170–171, 218–225

rhythms and, 103–104
temperature and, 90, 91–93, 95–97
Adaptation index, 162, 170
Adenine, 52, 55
Adenosine triphosphate
  ethionine and, 55
  synthesis of, 120
S-Adenosylethionine, 55
S-Adenosylmethionine, 55
Adrenal cortical secretion, 171, 172
Adrenal cortical stimulation, fasting and, 222
Adrenalectomy, 61
Adrenals, 159, 219
  tumors of, 67
Adrenocorticotropic hormone, 71
  carcinogenesis and, 67
  reserpine and, 23, 48
Aerosols, absorption of, 9
Aflatoxins, 220
  enzyme-forming systems and, 52–53
Age
  death rate and, 107
  toxicity and, 36
Age pigment, 119–120, 122, 140
Aggression, 207
  heat and, 188–190
Aging
  definition of, 105, 111
  health and, 208
  hypothesized causes of, 137–141
  model of, 132–133
  physiological function and, 147–148
  rate of
    biological factors and, 122–128
    chemical sources of disruptive changes and, 117–122
    kinetic energy transfer and, 115–117
    modification of, 128–136
    radiant energy and, 113–115
ALA synthetase, 46
Alanine, 60, 135
Alanine mustard, 38
Alcohol, 163
  absorption of, 9
Allergy, 46, 47, 163
  spontaneous regression of cancer and, 61
Altitude
  acclimation to, 159–162, 163, 164

adaptation to, 151, 171
cross-adaptation and, 173
Aluminum oxide, bronchial response to, 81
Alveolar wall
  irritants and, 78
  structure of, 79
Amethopterin, 38
Amines
  absorption of, 8
  red cell penetration by, 9
Amino acids, 162
  absorption of, 8
  central nervous system penetration by, 11
  membrane penetration by, 13
δ-Aminolevulinic acid, 46
Aminolevulinic acid synthetase, 46
Aminopyrine
  central nervous system penetration by, 10, 11
  metabolism, phenylbutazone and, 32
Ammonia, bronchial response to, 80–81, 82
Amphetamine, accumulative pattern and, 26
Androgen, carcinogenesis and, 69
Androsterone, 125
Anemia, hemolytic, 46
Aniline, 46
  central nervous system penetration by, 10, 11
  red cell penetration by, 9
Antibiotics, 53, 163, see also specific substances
  toxicity and, 38, 40–41
Antibodies
  adaptation and, 152
  antineural, 124
Anticholinesterases, 13
Anticoagulants, 29–31, 32, 33
Antigens
  autoimmune reaction and, 124
  neoplasms and, 72–73
Antihistamines, 163
Antimetabolites, toxicity and, 38, 39, 40
Antineural antibodies, 124
Antipyrine
  central nervous system penetration by, 11
  metabolism
    chlordane and, 32–33, 34
    phenobarbital and, 32
Apathy, heat and, 188, 189
Arabinosides, 123
Ascorbic acid, 171, 172
  adrenal, reserpine and, 23
  liver, 34, 171
*Aspergillus flavus*, aflatoxin synthesis and, 52
ATP, see Adenosine triphosphate

Atropine, 80
  bronchoconstriction and, 81
Autoimmunity, 124
  aging and, 139
Auxin, aging and, 124
Azo dyes, carcinogenesis and, 71

B

Barbital, central nervous system penetration by, 11
Barbiturates, see also specific compounds
  absorption of, 8
  pressor effect of paraoxon and, 19
Barometric pressure, adaptation to, see Altitude
BCNU, 38
Beans, monoamine oxidase inhibitors and, 25–26
Behavior, 162–163
  adaptation and, 93, 171
  heat and, 188–190
Behavioral codes, 155–156
3,4-Benzpyrene, 63
Benzpyrene hydroxylase, 219
Bishydroxycoumarin, see Dicumarol
Blood, see also Plasma; Platelets; Red Cells
  metabolism and, 15
  volume of, 172
Blood-CSF boundary, 11
Blood pressure, 162, 171, 172–173, 210
  dopamine and, 24–26
  paraoxon and, 19
Body temperature, 162, 171, 172
  adaptation to heat and, 178–183, 184–188, 189, 194–195, 200–201, 203
  aging and, 130
  comfort and, 157
  cross-adaptation and, 159, 163
  regulation of, 212–213
Body weight, 173
Brain, see also Central nervous system
  changes in amine level of, 21–24
5-Bromouracil, absorption of, 8
Bronchial responses, 80–84
Bronchitis, 80, 81
Bronchoconstriction, 80, 81–83
Bronchopneumonia, 80
Bufotenin, red cell penetration by, 9

C

Caffeine, 163
Calcium, adaptation and, 152
Calcium carbonate, bronchial response to, 81
Cancer, see Carcinogenesis; Neoplasm
Carbachol, bronchial response to, 81

Carbohydrate
  cardiac disease and, 209
  liver glycogen and, 220–221
  restriction, aging and, 129
Carbon monoxide, accumulative pattern of, 16
Carbon tetrachloride, 45
  enzyme-formation systems and, 54–55
Carcinogenesis, 46, 219, 224
  aflatoxins and, 52–53
  chemical, 67
  deoxyribonucleic acid and, 49, 50, 220
  dose-response relationships in, 62, 67
  ethionine and, 55
  hormonal factors in, 66–71
  immunologic factors in, 72–73
  nitrosamines and, 62, 74–75
  nutritional factors in, 73–74
  two-stage theory of, 62–64
  virus and, 123
Cardiac disease, 206–210
Catecholamines, 162, 171
  accumulative pattern of, 24
  cross-adaptation and, 164
  reserpine and, 48
Central nervous system, 164
  adaptability of, 149
  cardiac disease and, 210
  penetration of, 10–12
Cerebrospinal fluid, penetration into, 10–12
Cheese, monoamine oxidase inhibitors and, 25–26
Children, urban, 207
Chlorcyclizine, steroid metabolism and, 33
Chlordane
  antipyrine metabolism and, 32–33, 34
  hexobarbital metabolism and, 32
  phenylbutazone metabolism and, 32–33
  protein synthesis and, 57
  steroid metabolism and, 57–58
  warfarin metabolism and, 31, 33
Chlorpromazine
  accumulative pattern of, 20–22, 26
  benzpyrene hydroxylase and, 219
Cholesterol, biosynthesis of, 220
Choline, 55
Cholinesterase, parathion and, 17–20
Choroiocarcinoma, 61
Chromosomes, see also Genetics
  aging and, 148
Circadian rhythm, 103–104
Climate, 163
  heat adaptation and, 183–188, 192, see also under Adaptation
*Clostridium botulinum*, 8

Coal dust, bronchial response to, 81
Coenzymes, adaptation and, 94
Cold, see also Temperature
  adaptation to, 202–203
  aging and, 116–117
  reserpine and, 23
  tolerance, 150–151, 152
Collagen, aging and, 125, 130, 140, 141
Colloidal silica, 80
Colon, see Gastrointestinal tract
Conservation, definition of, 214
Corticosteroids, 33, 125
  carcinogenesis and, 70, 71
  feeding habits and, 221, 222
  hydroxylation of, 58
  metabolism, diphenylhydantoin and, 33–34
  neoplasm and growth of, 71
  paraoxon and, 20
  reserpine and, 23
Cough, 80, 81, 83
Creatine, 161
Creosote, 64
Cross-adaptation, 158–165, 170–175
  acceleration and, 162
  altitude and, 159–162, 163, 164
  definition of, 158
  hyperoxia and, 163
  hypoxia and, 162, 163
  methods for study of, 161–162, 172–175
  nutrition and, 160–161, 163
  temperature and, 159–161, 162, 163–165
  tolerance to infection and, 163
Culture, 216–217, 224
  adaptation and, 149–153, 154
Cytochrome C, 60
Cytosine, 52
Cytoxan, 38

## D

$o,p'$-DDD, 173–175
DDT, 174
  accumulation pattern of, 16, 26
  hexobarbital metabolism and, 32
  toxicity of carbon tetrachloride and, 54
  warfarin metabolism and, 31, 33
Death, heat and, see Heat stroke
Death rate,
  age and, 107
  exposure to heat and, 109–110, 111
  exposure to nonlethal harmful substances and, 109–110
  infection and, 110
  ionizing radiation and, 110, 112
Decamethonium, central nervous system penetration by, 11

Dehydration, 181
Deoxyribonucleic acid
  aflatoxin and, 53
  aging and, 123, 134, 137, 148
  carcinogenesis and, 49, 50
  cardiac disease and, 210
  damage to, 220
    X-ray, 115
  protein synthesis and, 51–52
Desoxycorticosterone metabolism, drug stimulation of, 33
Deuterium, life expectancy and, 119, 120
Development, 148
  life expectancy and, 121–122, 129
Dibenzyline, 47
Dicumarol metabolism, phenobarbital and, 29–31
Diet, see also Nutrition
  cold tolerance and, 151, 152
  enzyme adaptation and, 220–224
  vitamin E deficiency in, 119–120, 122
Diffusion, membrane penetration by, 6–7
Dilantin
  cortisol metabolism and, 33–34
  metabolism of, phenobarbital and, 32
  steroid hydroxylation and, 58
Dimethylbenzanthracene, 219
Dimethylnitrosamine, 47, 75
  carcinogenesis and, 50
Diphenylhydantoin, see Dilantin
Dipyrone metabolism, barbiturates and, 32
Diuretics, 163
DNA, see Deoxyribonucleic acid
Dopa, beans and, 26
Dopamine, accumulative pattern of, 24–26
Drugs, 181
  absorption of from gastrointestinal tract, 8–9
  accumulation of, 163
  adaptation to, 171
  aging and, 129, 132–136
  allergic response to, 47
  analgesic, 28, 32
  antibiotic, see Antibiotics
  anticoagulant, 29–31, 32, 33
  anticonvulsant, 32
  antidepressant, 48
  antihistaminic, 163
  antimalarial, 11, 46
  antineoplastic, toxicity and, 38–39, 40
  antirheumatic, 47
  cross-adaptation and, 163–164, 173–175
  development of, toxicity and, 41–42
  hypnotic, 28, 32
  metabolism of, 57–58, 173–175

  species differences in, 37
  stimulation of, 28–34
  toxicology and, 44
  purity of, toxicology and, 44
  radiometric, 131
  sympatholytic, 163
  sympathomimetic, 163
  toxicity of, see Toxicity
  tranquilizing, 219
Dust, bronchial response to, 81, 82
Dyspnea, 80

E

Ear-duct, carcinogenesis and, 71
EEG, see Electroencephalogram
Electroencephalogram, heat and, 189
Electrolytes, membrane penetration by, 8
Enterohepatic circulation, 12
Environment, optimum, 216–225
Environmental agents, definition of, 16
Environmental physiology, 89–90
Enzyme-forming systems
  aflatoxins and, 52–53
  carbon tetrachloride and, 54–55
  ethionine and, 55–56
Enzymes, see also specific substances
  adaptation and, 90, 93, 94, 95, 152, 219–224
  aging and, 119, 121, 135–136, 137, 140, 141
  cross-adaptation and, 173–175
  deficiencies and, 46
  drug metabolizing, 28–34, 44
  induction of, 28–34
  liver, 15, 28–34
  peroxyacetyl nitrate and, 59
  synthesis of, 219
Epilepsy, 189
Epinephrine, 55, 125, 141
  red cell penetration by, 9
Ergonomics, 205
Erythrocyte, penetration of, 10
Eserine, 80
Estradiol, 57
Estradiol-17$\beta$ metabolism, drug stimulation of, 33
Estrogens, carcinogenesis and, 67, 69–70
Ethanol, circadian rhythm and, 103
Ethionine
  cross-adaptation and, 174–175
  enzyme-forming systems and, 54, 55–56
  liver and, 77
Ethylcarbamate, 47
Ethylene, auxin and, 124

# INDEX

Evolution, 88, 101, 153, 218, see also Selection
  biological rhythms and, 103
  harmful agents and, 112
Excretion, species differences in, 37
Exercise, see Physical activity

## F

Fasting, enzyme adaptation and, 221–224
Fat
  absorption of, 8
  cardiac disease and, 209
  fasting and, 221
Fatty acids
  aging and, 125
  free, 141
    paraoxon and, 20
Fatty liver, 54
  ethionine and, 55
Fetus, drugs and, 47
Fever
  aging and, 118
  spontaneous regression of cancer and, 61
Fluorenylacetamide, see Acetylaminofluorene
5-Fluorodeoxyuridine, 38
5-Fluorouracil, 38
  absorption of, 8
Food, see also Nutrition
  effect of dopamine on consumption of, 25–26
Food intake, distribution of, 220–224
5-FUDR, 38

## G

Gastrointestinal tract, see also Intestine
  membrane penetration in, 8–9, 12, 14
Genetic damage, 115, 220,
  aging and, 137
Genetics, 156, 171, 217, 225, see also Mutation
  adaptation and, 88, 89, 90, 93, 95, 98
  carcinogenesis and, 49, 50, 67, 72
  cardiac disease and, 206, 209–210
  toxicity and, 46
Gluconeogenesis, 222
Glucorinide, 219–220
Glucose
  central nervous system penetration by, 11
  fasting and, 221, 222
  metabolism, drug stimulation of, 34
  paraoxon and, 20
  tolerance, 125
Glucose-6-phosphatase, liver carcinogenesis and, 77

Glucose-6-phosphate, 59
  red cells and, 46
Glucose 6-phosphate dehydrogenase
  feeding habits and, 220–221, 223
  peroxyacetyl nitrate and, 59
Glucuronic acid, membrane penetration and, 12
Glutamic acid, 60
Glutamic dehydrogenase, 60
Glutathione, 46
  peroxyacetyl nitrate and, 59
Glutethimide, drug metabolism and, 32
Glycine, 55
Glycogen, 125, 222
  feeding habits and, 220–222
  liver carcinogenesis and, 77
Goitrogens, carcinogenesis and, 68–69
Gonadectomy, carcinogenesis and, 67
Gonadotropins, 71, 125
  carcinogenesis and, 67
Gonads, carcinogenesis and, 69–70
Griseofulvin metabolism, phenobarbital and, 32
Growth, see also Development
  fasting and, 223
  neoplasms and, 65, 71
  stimulation of, carcinogenesis and, 67
Growth hormone, aging and, 139
Guanine, 52
  $\beta$-propriolactone and, 49

## H

Hair follicles, absorption through, 9
Health, 218–219, 224
  definition of, 155
Heart, see also Cardiac disease
  weight of, 171
Heart rate, 162
  adaptation to heart and, 184–187, 194–195
  monoamine oxidase inhibitors and, 25
  paraoxon and, 19
Heat, see also Temperature
  adaptation to, see Adaptation
  aging and, 118–119, 140
    rate of, 115–116
  comfort and, 157
  life expectancy and, 109–110, 111
Heat stroke, 178–180, 182–183, 200–202
Heavy water, see Deuterium
Hemoconcentration, 171, 172
Hemoglobin, peroxyacetyl nitrate and, 59
Hemolysis, 46
Hemolytic anemia, 46

Hepatomas
  aflatoxin and, 220
  feeding habits and, 220–221
Heptabarbital, coumarin metabolism and, 31
Heritability, 101, see also Genetics
Hexamethonium, central nervous system penetration by, 11
Hexobarbital, metabolism of, 174–175,
  chlordane and DDT and, 32
  phenobarbital and, 32
Hippuric acid, red cell penetration by, 10
Histamine, bronchial response and, 82, 83
Histones, 137
Homeostasis, 89, 90–91, 156
Homeotherms, 95–96, 129, 130, 212
Homograft reaction, 72
Hormones, see also specific substances
  aging and, 124–125, 139
  carcinogenesis and, 66–71
  cross-adaptation and, 163, 164
  mammotropic, see Mammotropic hormone
  neoplasms induced by, 64, 65
  neoplasms treated with, 65
  ovarian, carcinogenesis and, 69
  pituitary, 20
17-Hydroxycorticosteroids, 162, 164
6-Hydroxycortisol, 34
6$\beta$-Hydroxycortisol, 58
N-Hydroxyfluorenylacetamide, 71
Hydroxylamines, toxicity of, 46
Hydroxyurea, 38
Hyperbilirubinemia, 219
Hyperglycemia, 20
Hypernephroma, 61
Hyperoxia, cross-adaptation and, 163
Hypertension, 209
  cross-adaptation and, 172–173
Hypophysectomy, 61
  carcinogenesis and, 71
  mammary tumors and, 70
Hypothalamus, 213
Hypoxia, cross-adaptation and, 162, 163, 172–173
Hysteria, heat and, 188, 189

I

Ice, aging and, 116–117
Illness, 218–219, 224
  definition of, 155
  heart and, see Cardiac disease
  heat and, 192–193, 202, see also Heat stroke
  psychosomatic, 192
  urban residents and, 207
Imipramine, monoamine oxidase inhibitors and, 48

Immunity, 61 see also Autoimmunity
  carcinogenesis and, 72–73
Imprinting, autoimmune reaction and, 124
Indole, 222
Infection
  adaptation and, 152
  aging and, 123
  life expectancy and, 110
  spontaneous regression of cancer and, 61
  tolerance to, 163, 164
Injury, aging and, 116
Inosine, 55
Instinct, 221
Insulin, 164
  carcinogenesis and, 67
Intestine, see also Gastrointestinal tract
  enzyme adaptation and, 219
  metabolism and, 15
  weight of, 171
Ionization
  central nervous system penetration and, 10–11
  damaging, 114
Ionizing radiation, see Irradiation
Ions
  absorption of, 8
  adaptation and, 94
Irradiation, 171
  aging and, 130–132, 133, 140, 148
  carcinogenesis and, 67, 70
  circadian rhythm and, 103
  life expectancy and, 110, 112, 129
Isocitrate, 59
Isocitrate dehydrogenase, peroxyacetyl nitrate and, 59
Isoniazid, toxicity of, 45
Isoproterenol, bronchoconstriction and, 82
Israunescine, 22
Isoreserpine, 22
Isozymes, adaptation and, 94

K

17-Ketosteroids, 171, 172
Kidney, 173
  carbon tetrachloride and, 54
  membrane penetration and, 7, 12
  neoplasms of, 67
  weight of, 171, 172

L

Lasiocarpine, 53
Leptospirosis, 41
Leukemia, 67, 73
Life expectancy, 119–122, 129, 155
  fasting and, 223

Ligase, 135, 136
Light
  efficiency of in producing damage, 114
  infrared, aging and, 113
  ultraviolet, aging and, 114
  visible, aging and, 113
Lipids
  adaptation and, 94
  age pigment and, 120
  solubility of
    membrane penetration and, 7–14
    toxicity and, 46
Lipofuscin, see Age pigment
Lipoperoxidation, 54
Liver
  carbon tetrachloride and, 54
  carcinogenesis and, 71, 77
    nutrition and, 74
  cortisol hydroxylation in, 58
  drug metabolism and, 28–34
    cross-acclimation and, 173–175
  enzyme adaptation and, 219–222
  growth of, 34
  membrane penetration and, 12
  metabolism and, 15
  neoplasms of, 52–53, 55
  ovarian neoplasms and, 69
  paraoxon and, 20
  protein synthesis in, 51–56, 57
  thyroxine metabolism and, 175
  triglyceride accumulation in, 54
  weight of, 171, 172, 173, 174
Lungs, see also Alveolar wall
  bronchial responses to inhalants and, 80–84
  nerves in, 82
  structural alterations of, 78
  thyroid tumors and, 69
  volume of, 81
Lupus erythematosis, 124
Lymphatics, thyroid neoplasms and, 69

## M

Magnesium, 161
Malate dehydrogenase, peroxyacetyl nitrate and, 59
Mammary gland
  carcinogenesis and, 70
  neoplasms of, 61, 67, 111
Mammotropic hormone, carcinogenesis and, 67, 70
Mannitol, absorption of, 8
Mebutamate, pressor effect of paraoxon and, 19
Mecamylamine, central nervous system penetration by, 11

Melanin, ultraviolet light and, 114
Melanoma, malignant, 61
Melatonin, 55
Membrane penetration, 6–14
Mental disturbance, heat and, 190–193
Meprobamate, drug metabolism and, 32
6-Mercaptopurine, 38
Metabolism, 119, 164
  adaptation and, 91, 95, 96–97
  alteration of, 52
  daily routine and, 221–223
  drugs and, see Drug metabolism
  heat and, 182, 203
  impairment of, toxicity and, 46
  phases of, 15
  protein and, 151, 152
  stimulation of, 28–34
  thyroxine and, 175
Metastasis, neoplasm and, 65
Methemoglobinemia, 46
Methionine, ethionine effects and, 55
Methotrexate, 38
Methylcholanthrene, metabolism and, 52
3-Methylcholanthrene
  mammary tumors and, 70
  protein synthesis and, 57
Methyl-GAG, 38
$N^1$-Methylnicotinamide, central nervous system penetration by, 11–12
Methyl seratonin, see Melatonin
Mitochondria, aging and, 138
Mitomycin C, 38
Monoamine oxidase inhibitors
  accumulative pattern of, 24–26
  drug interaction and, 48
Monosaccharides, membrane penetration by, 7, 8
Morphine, cross-adaptation and, 172, 173
Mortality, model of, 107–108
Motivation, 151–152
  heat stress and, 193, 195
Mucus, bronchial irritants and, 83
Muscles, 213
Mutation, 101, 218
  aging and, 115, 137, 148
  autoimmune reaction and, 124
Mycotoxin, 52
Myleran, 38

## N

Naphthylamine, 46
Necrosis, liver-cell, 53, 54, 55, 56
Neoplasm, 68, 73
  benign, 62–63
  cross-adaptation and, 172, 173

dependent and autonomous, 66
formation of, 49, 50
mammary, 61, 67, 70
  life expectancy and, 111
metastasis and, 65
nutrition of, 61
ovarian, 67, 69
progression of, 64–66, 71
spontaneous regression of, 61–62
surgical removal of, 61
testicular, 69
uterine, 67
Neoprontosil, 60
Neotetrazolium, 60
Nerves, respiratory tract and, 81, 82
Neuroblastoma, 61
Nialamide
  accumulative pattern and, 26
  food consumption and, 25–26
Nicotinamide, 171
Nicotine, metabolism of, 16
Nitrogen mustards, 13, 38
Nitromin, 38
5-Nitrosalicylic acid, central nervous system penetration by, 11
Nitrosamines, carcinogenesis and, 62, 74–75
N-Nitrosomethylurea, carcinogenesis and, 50
Norepinephrine, 26, 48
  red cell penetration by, 9
  reserpine and, 22
Nutrition, 156, 171, 172, see also Diet; Food
  adaptation and, 152, 182
  aging and, 119–120, 121–122, 124, 125, 128, 129
  carcinogenesis and, 73–74
  chemical constituents in blood and, 157
  cross-adaptation and, 160–161, 163
  health and, 208, 209
  of neoplasm, 61
Nuts, pyrrolizidine alkaloids in, 47

## O

Obesity
  feeding habits and, 222–223
  life expectancy and, 122
Oophorectomy, 61
Ovariectomy, mammary tumors and, 70
Ovary, tumors of, 67, 69
Oxygen consumption, adaptation and, 95–96

## P

Pamaquine, toxicity and, 46
Pancreas, carbon tetrachloride and, 54
Pancreatitis, 55
Pantothenate, 122

Papillomas, 62–63
Paraoxon
  accumulative pattern and, 26
  production of from parathion, 19
Parasympathetic nervous system, paraoxon and, 19
Parathion, 15
  accumulation pattern of, 17–20
Pargyline, food consumption and, 25
Peanuts, aflatoxins and, 53
Penicillin, 47
  central nervous system penetration by, 11
  toxicity of, 40–41
Pentobarbital
  central nervous system penetration by, 11
  circadian rhythm and, 103
  metabolism of, 173–175
Pentose, 171
Peroxyacetyl nitrate, effects of at enzyme level, 59
Personality, heat and, 191
Pesticides, see also specific compounds
  accumulation patterns of, 16–20, 26
  development of, 41
Phagocytosis, 7, 13
Phenobarbital
  antipyrine metabolism and, 32
  Dicumarol metabolism and, 29–31
  dilantin metabolism and, 32
  enzyme adaptation and, 219–220
  griseofulvin metabolism and, 32
  hexobarbital metabolism and, 32
  metabolism and, 52
  protein synthesis and, 57
  steroid metabolism and, 33–34, 57–58
  warfarin metabolism and, 29, 31, 33
  zoxazolamine metabolism and, 28–29, 30
Phenolic-indolic acids, 162
Phenol red, red cell penetration by, 10
Phenothiazine, 219
Phenotypic plasticity, 149
L-Phenylalanine mustard, 38
Phenylbutazone
  aminopyrine metabolism and, 32
  metabolism, chlordane and, 32–33
  steroid hydroxylation and, 58
  steroid metabolism and, 33
Phenylhydroxylamine, toxicity and, 46
Phosgene, bronchial response to, 80–81
Phosphorus, 161
Photosynthesis, 113
Physical activity
  cross-adaptation and, 172, 173
  fasting and, 221–223
  health and, 208–209

Pinocytosis, 7, 9, 13
Pituitary, 159
　aging and, 139
　growth hormone of, 67
　neoplasms of, 67
Pituitary-gonadal system, carcinogenesis and, 69–70
Pituitary hormones, see specific hormones
Pituitary-mammary gland system, carcinogenesis and, 70
Pituitary-thyroid system, carcinogenesis and, hormones and, 68–69
Plant alkaloids, absorption of, 8
Plasma, 10, 11, 12, 13
　volume of, 171, 172
Plastids, aging and, 138
Platelets, penetration of, 9–10
Pneumonia, 78
　bronchial, 80
Poikilosmotic animals, adaptation and, 90
Poikilothermic animals, 129
　adaptation and, 90, 91–93, 95–96
Pollution, definition of, 214
Polymorphism, 98
Polyoma virus tumors, 73
Polysomes, 52
Pores, membrane penetration and, 7, 10, 12, 13
Porphyria, 46
Prediction, toxicity and, 37–42
Primaquine, toxicity and, 46
Procaine amide, red cell penetration by, 9
Productivity, 225
　heat and, 195–200
Progesterone, 57–58
　carcinogenesis and, 67
　metabolism, drug stimulation of, 33
Prolactin, 71
$\beta$-Propriolactone, deoxyribonucleic acid and, 49
Protein
　cold tolerance and, 151, 152
　denaturation of, 116
　　aging and, 118–119
　gluconeogenesis from, 222
　membrane penetration and, 7, 8
　restriction of, 129
　structure of, 93–94
　synthesis of, 220–224
　　adaptation and, 94
　　aging and, 138
　　$o,p'$-DDD and, 174–175
　　deoxyribonucleic acid and, 51–52
　　ethionine and, 174–175
Protozoa, cross-adaptation and, 172, 173

Psittacosis virus, 152
Psychoneurosis, 192, 193
Pulmonary edema, 78
Pulmonary membrane, penetration of, 9
Purine, 52
Puromycin, 56
Pyridoxal phosphate, toxicity and, 45
Pyrimidines, 52
　absorption of, 8
Pyrrolizidine alkaloids, 47

Q

Quinine, central nervous system penetration by, 11

R

Race, adaptation and, 95
　heat and, 183–188, 190
Radiation, see Irradiation
Raunescine, 22
Red cells
　antimalarial agents and, 46
　glucose-6-phosphate deficiency in, 46
　penetration of, 9–10, 13
Reductases, 59, 60, 75
Reflex
　adaptation and, 93
　bronchoconstriction and, 82
Rescinnamine, 22
Reserpine, 48
　accumulative pattern and, 20, 22–24, 26
Respiratory tract, see also Alveolar wall; Lungs
　absorption from, 9
　membrane penetration of, 14
Response, adaptation and, 88
Rheumatoid factor, 124
Rhythms, adaptation and, 103–104
Ribonuclease, 59
Ribonucleic acid, 55, 56
　aging and, 133–138
　carcinogenesis, 49
　enzyme formation and, 51–52
　synthesis, adaptation and, 94
　　aging and, 123
RNA, see Ribonucleic acid

S

Salicylates,
　absorption of, 8
　central nervous system penetration by, 11
Saliva, 162
Sarcomas, 73
Sedation, drug accumulation patterns and, 21, 22

Selection, 155, 218
　aging and, 125
　cultural adaptation and, 150
　feeding habits and, 222–223
　genetic adaptation and, 88, 101–102
　resistance to harmful agents and, 112
　temperature adaptation and, 150
Senescence, see Aging
Serotonin, 222
　accumulative pattern of, 24
　platelet penetration by, 10
　red cell penetration by, 9
Serpentine, 22
Shivering, 203, 213
Skin, 71
　absorption from, 9
　disease of, heat and, 192, 193
　drugs and, 47
　penetration of, 14
　temperature adaptation and, 90
　temperature of, 213
　ultraviolet light and, 114
Sleep, 162
Smog, bronchial response to, 81
Smoke, 16
　bronchial response to, 81, 82
Sodium, 161
Somatotropin, carcinogenesis and, 70, 71
Space, competition for, 126, 128
Species
　definition of, 36, 89
　extrapolation of toxicity data and, 36–42
Spleen, ovarian tumors and, 69
Steroids, 221
　metabolism, drugs and, 33–34, 57–58
Stomach, see Gastrointestinal tract
Stress, biological rhythms and, 104
Sugars
　absorption of, 8
　membrane penetration by, 13
Sulfaguanidine
　absorption of, 8
　central nervous system penetration by, 11
Sulfanilic acid, red cell penetration by, 10
Sulfonic acid dyes, central nervous system penetration by, 11
Sulfonic acids, absorption of, 8
5-Sulfosalicylic acid, central nervous system penetration by, 11
Sulfur dioxide, bronchial response to, 81, 82
Sulfuric acid, bronchial response to, 81
Sweating, 213
　heat adaptation and, 181–182, 184–188
Sympathetic nervous system, drug interaction and, 48

Sympatholytic drugs, 163
Sympathomimetic drugs, 163

T

Tachypnea, 80, 83
Tactile stimulation, bronchial response to, 82
Temperature, see also Cold; Heat
　adaptation and, 90, 91–93, 95–97, see also under Adaptation
　　cultural, 150–151, 152
　body, see Body temperature
　cross-adaptation and, 159–165, 170–173
Testis, 171
　neoplasms of, 67, 69
　weight of, 172
Testosterone, 57
　metabolism, drug stimulation of, 33
Tetramethylammonium, red cell penetration by, 9
Thalidomide, 42, 47
Thermoregulation, 212–213
Thioesterase, 59
Thiopental, central nervous system penetration by, 10, 11
ThioTEPA, 38
Thiouracil, carcinogenesis and, 68–69
Threshold, carcinogenesis and, 62
Thymine, absorption of, 8
Thyroid, 159, 171, 172
　neoplasms of, 67, 68–69
　weight of, 175
Thyroidectomy, carcinogenesis and, 71
Thyroid hormone, carcinogenesis and, 68–69
Thyrothropin, carcinogenesis and, 67
Thyroxine
　carcinogenesis and, 67
　metabolism of, 175
Thyroxine-$^{131}$I, 175
Tobacco, 163
Tobacco smoke, contents of, 16
Tolerance, see Cross-adaptation
Toxicity, 15, 44–48
　accumulation patterns and, 16, 17, 26
　extrapolation from animals to man and, 36–42, 44
　species and, 52
Toxins
　aging and, 124
　bacterial, 171
　　cross-adaptation and, 172, 173
　enzyme adaptation and, 219–220
　life expectancy and, 109–110
TPNH, see Triphosphopyridine nucleotide
Tranquilizers, 219
Trichinosis, 152

Triglyceride, liver and, 54, 55, 56
Triphosphopyridine nucleotide, 46
Triphosphopyridine nucleotide cytochrome C reductase, 60
5-OH-Tryptophan, 222
Tryptophan-pyrrolase
  paraoxon and, 20
  reserpine and, 23
Tumor, see Neoplasm
Turpentine, life expectancy and, 109–110
Tyramine, nialamide and, 26
Tyrosine transaminase
  feeding habits and, 221, 222
  paraoxon and, 20
  reserpine and, 23

## U

Ultraviolet light, see Light
Uracil, 52
  absorption of, 8
Urea, 161
  derivatives, absorption of, 9
Urethane, 47
Uric acid, 161
Urine
  adaptation and, 161, 162, 165
  6$\beta$-hydroxycortisol in, 58
Uterus, tumors of, 67

## V

Vagosympathetic nerves, bronchoconstriction and, 81
Vasoconstriction, 171
Vinblastine, 38
Vincristine, 38
Virus, 73, 152
  aging and, 123, 139
Vitamin E, age pigment and, 119–120, 122

## W

Warfarin metabolism
  insecticides and, 31, 33
  phenobarbital and, 29, 31, 33
Wastage, 214
Weightlessness, 164, 171
Weight loss, cross-adaptation and, 160–161
Wind velocity, heat and, 193–202

## X

X-rays
  aging and, 130–132, 133
  cross-adaptation and, 172, 173
  damage produced by, 114–115

## Z

Zoxazolamine metabolism, phenobarbital and 3,4-benzpyrene and, 28–29, 30

PRINCIPAL,
REGIONAL TECHNICAL COLLEGE,
PORT ROAD,
LETTERKENNY.